Fluid Dynamics and Computational Modeling

Volume II

Fluid Dynamics and Computational Modeling Volume II

Edited by **Maria Forest**

CLANRYE INTERNATIONAL

New Jersey

Published by Clanrye International,
55 Van Reypen Street,
Jersey City, NJ 07306, USA
www.clanryeinternational.com

Fluid Dynamics and Computational Modeling
Volume II
Edited by Maria Forest

International Standard Book Number: 978-1-63240-239-4 (Hardback)

Printed in the United States of America.

Contents

Preface

The main aim of this book is to educate learners and enhance their research focus by presenting diverse topics covering this vast field. This is an advanced book which compiles significant studies by distinguished experts in the area of analysis. This book addresses successive solutions to the challenges arising in the area of application, along with it; the book provides scope for future developments.

This book contains updated information on topics like fluid dynamics, computational modeling and its applications. It discusses topics like: winds, building and risk prevention; multiphase flow, structures and gases; heat transfer, combustion and energy; medical and biomechanical applications; and other crucial topics. Along with all that has been mentioned, this book also gives a detailed view of computational fluid dynamics and applications, without excluding experimental and theoretical aspects.

It was a great honour to edit this book, though there were challenges, as it involved a lot of communication and networking between me and the editorial team. However, the end result was this all-inclusive book covering diverse themes in the field.

Finally, it is important to acknowledge the efforts of the contributors for their excellent chapters, through which a wide variety of issues have been addressed. I would also like to thank my colleagues for their valuable feedback during the making of this book.

Editor

Part 1

Medical and Biomechanical Applications

Surfactant Analysis of Thin Liquid Film in the Human Trachea via Application of Volume of Fluid (VOF)

Sujudran Balachandran
Bumi Armada Berhad
Malaysia/Singapore

1. Introduction

The human tracheobronchial tree is a complex branched distribution system in charge of renewing the air inside the acini which are the gas exchange units. The surfactant factor existing in the acini of the human tracheobronchial tree is exposed to thin liquid film flow where the application of volume of fluid (VOF) can significantly determine the membrane effects on the branching asymmetry. The thin film application in this chapter will focus on the breathing airway which is commonly known as the windpipe (trachea) of an adult human. The upper human airway is the primary conduit for inspiration in the breathing process. Air entering the mouth passes through pharynx and flows into the trachea via the glottal region. The air which enters the windpipe applies a surfactant pressure at the lipo-protein complex, where this complex is developed as a thin layer on the trachea (Caro et al. 2002). The surfactant is a lipo-protein complex, which is a highly surface-active material found in the fluid lining of the air-liquid interface in the trachea surface. This surfactant plays a dual function of preventing sudden collapse during the breathing cycle and protection from injuries and infections caused by foreign bodies and pathogens. The varying degrees of structure-function abnormalities of surfactant have been associated with obstructive trachea diseases, respiratory infections, respiratory distress syndromes, interstitial lung diseases, pulmonary alveolar proteinosis, cardiopulmonary bypass surgery and smoking habits. For some of the pulmonary conditions, especially respiratory distress syndrome, surfactant therapy is on the horizon. In order to understand the behaviour and relevant condition of the surfactant in the human trachea, it is important to apply the volume of fluid method on these surfaces. The phenomena that occur on the trachea will ensure that the surfactant responsibility in resolving the potential obstruction of breathing. The surfactant factor may occur with non-lateral conditions in space as well as during the inspiration of breathing in the human body. The fluid interaction in the thin surface results in serious impairment by obstructive trachea diseases as mentioned earlier. The pulmonary surfactant is essential for normal breathing, alveolar stability and as a host defence system in the lungs. The interface of surfactant films reduces the surface tension to extremely low values when it is compressed during expiration. This protects our lungs from collapse during breathing out. Thus application of the Volume of Fluid (VOF) method is introduced in this paper to study the behaviour of the pulmonary surfactant in the human trachea.

2. Model selection

Figure 1 shows the three-dimensional model of the lung. Detailed geometries of this model were extracted from the anatomical model by Schmidt et al. (2004). This explicit human lung was derived from High-Resolution Computed Tomography (HRCT) imaging of an in vitro preparation. This model was a selection of data extracted from the anatomy of a healthy human lung of an adult male. This lung model is free from pathological alteration (Gemci et al. 2008). In Figure 2, an isometric view of the human lung is presented. The notation of G0, G1, G2 and G3 are the cross-sectional gate to facilitate the model drawn for indication segments. This model was selected based on the actual structure of the human lung. No additional tissue configuration was created due to investigation which only focused on the airway blockage area. At each generation, the branching is essentially dichotomous; each airway being divided into two smaller daughter airways (Farkas et al. 2007). The tree starts at the trachea (G0-G0) whose average diameter and length are respectively D0 = 1.8 cm and L0 = 12 cm in the healthy human adult (Allen et al. 2004), and ends in the terminal bronchioles. From the trachea to terminal bronchioles, this element is located on average in generation G1-G1, these airways are purely conducting pipes. Two important features have

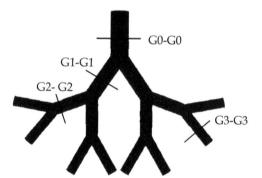

Fig. 1. General three-dimensional lung airway model

Fig. 2. Isometric view of lung airway

been included. First, the airway dimensions at the first generations (G0-G0 and G1-G1) are specific to the human anatomy and they are essentially independent of physiological variability (Martonen et al. 2003). Second, for higher generations (G2-G2 and G3-G3), a systematic branching asymmetry has been modelled in the different tree bifurcations.

Figure 3a and 3b illustrate the overall mesh of the model selected. The meshes are generated using a tetrahedral unstructured mesh format. The total number of grids generated is approximately 0.62 million. Concentrated mesh is applied near the branches of the bifurcation to increase the result interpretation. The flow details are critical as the effect of the boundary layer becomes significant in this region.

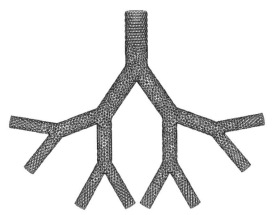

Fig. 3a. Mesh concentration on lung model

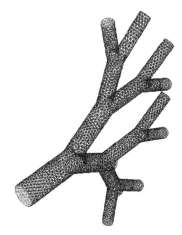

Fig. 3b. Mesh concentration on lung model (isometric view)

3. CFD simulation

The categorization of the anatomy of tracheobronchial tree airways constitutes the first step in the examination of the respiratory flow field and surfactant impact with volume of fluid

(VOF). The intention of this study is to understand the behaviour of the membrane which acts as a thin liquid film and the role of the surfactant on the model. In this study, the homogenous airflow is a comprehensive digital reference model with a maximum of four sections of pulmonary airway tree computed using the commercial CFD code FLUENT® (version 6.2). The computational mesh was generated using the FLUENT mesh generation code GAMBIT®. FLUENT employs a finite-volume method to solve the Navier-Stokes and continuity equations on an arbitrarily shaped flow domain with appropriate boundary conditions. The steady-state solution of the flow fields assumed converged when the residuals reduced to less than 10^{-4}. A typical run time was approximately 38 hours on a normal processor. In this CFD study of surfactant in the human pulmonary tree model, computations were performed at 28.3 L/min for a quasi steady-state volumetric adult inhalation flow rate for pulmonary surfactant.

3.1 Volume of Fluid (VOF) method in Fluent 6.2
The Volume of Fluid (VOF) model has been widely used in many research fields and commercial software. Known for its wide range of usage, implementation of this model in commercial code has been popular and FLUENT 6.2 is one of the software where their codes have been implemented. Isothermal VOF formulation was considered in this project with heat transfer and is assumed negligible due to analysis being based on surface interface. Formulation of laminar flow is presented here. For the turbulence effect, it can be done by averaging the Navier Stokes equation and detailed formulation can be found in Versteeg and Malalsekera (1995).

3.1.1 The basic of VOF for interface development
The VOF model is based on a volume fraction denoted by α where this volume of fraction is using a two fluids mixture in a fixed computational grid. The transport of the two fluids results in solving the multiphase fluid equation using the α values and interface location is reconstructed from the volume fraction field. The following procedures are used in solving the VOF model:
i.   Initial interface shape is modelled in computation domain.
ii. From the existing interface in the model, volume of fraction is obtained. (Truncated volumes calculation was done at each interfacial cell, this discrete volume fraction is used instead of the exact interface information.)
iii. The volume fraction field is adverted. (Transport of fluid.)
iv. Finally, the interface geometry and location are obtained using the new volume fraction.

All these steps are repeated until a convergence criterion is achieved. In VOF, especially in FLUENT 6.2, one can conduct a simulation for two different fluids without mixing the respective fluids. There will be an interface region between the both liquids and for each region α_q of 1 and 0 values will be given.

α_q = 0: the cell is empty
α_q = 1: the cell is full
$0 < \alpha_q < 1$: the cell contains the interface between the qth fluid and one or more other fluids.
Using the same momentum equation, both fluids shared the same velocity field and it is solved throughout their domain. The tracking interface between both phases is accomplished by the solution of a continuity equation for the volume fraction of one of the phases. The equation is:

$$\frac{\partial \alpha_q}{\partial t} + \vec{v} . \nabla \alpha_q = \frac{S_{\alpha_q}}{\rho_q} \tag{1}$$

The primary phase of the fluid will not be solved by the volume fraction equation, but will be computed through the following constrains:

$$\sum_{q=1}^{n} \alpha_q = 1 \tag{2}$$

For each of the phases, the properties will be presented in control volume of the respective component phase. For the two phase system, for example, if the phases are represented by the subscripts 1 and 2, and if the volume fraction of these is being tracked, the density in each cell is given by:

$$\rho = \alpha_2 \, \rho_2 + (1 - \alpha_2)\rho_1 , \tag{3}$$

and for every n – phase system , the volume fraction averaged density takes on the following form:

$$\rho = \sum \alpha_q \rho_q \tag{4}$$

3.1.2 The momentum equation
As mentioned earlier, both fluids will share the same momentum equation throughout the domain. The momentum equation is:

$$\frac{\partial}{\partial t} \rho u_j + \rho u_i \frac{\partial u_j}{\partial u_i} = -\frac{\partial P}{\partial x_j} + \mu \frac{\partial}{\partial x_i}\left(\frac{\partial u_i}{\partial x_j} + \frac{\partial u_j}{\partial x_i} \right) + \rho g_j + B_j , \tag{5}$$

Velocity component and direction denoted by u_j and x_j in j direction and t is the time. The summation of convention is applied when an index is repeated and i, j = 1, 2 for this problem. Pressure and gravitational acceleration are denoted by P and g respectively and B is a source term due to any additional force applied (per unit volume). This momentum equation dependant on the volume fraction α_q where in this case q = 1, 2 of the phases through the density ρ and viscosity μ . The computational domain is calculated through the following equation:

$$\frac{D\alpha_2}{Dt} = \frac{\partial \alpha_2}{\partial t} + u_i \frac{\partial \alpha_2}{\partial x_i} = 0 \tag{6}$$

The value of α_1 for the first phase is not solved directly at any stage, but it is solved based on the volume condition $\alpha_2 + \alpha_1 = 1$. By applying the properties of density ρ and viscosity μ , the following equations are obtained since ρ and μ depend on the volume fraction values:

$$\rho = \alpha_2 \rho_2 + \alpha_1 \rho_1 , \tag{7}$$

$$\mu = \alpha_2 \mu_2 + \alpha_1 \mu_1 \tag{8}$$

The volume transport equation then, yields an updated volume fraction field with discrete values in computational cell. The values will later be used in reconstructing the new interface location.

3.1.3 Interface reconstruction

Reconstruction of new interface needs formulation from the discrete values of volume fraction inside the domain. In FLUENT 6.2, there are four interface reconstruction schemes, which are geometric reconstruction, donor-acceptor, Euler explicit and Euler implicit. Both Euler explicit and Euler implicit schemes treat cells that lie near the interface using the same interpolation as other cells. Meanwhile, the geometric reconstruction and the donor acceptor scheme treat the interfaces in an advanced interpolation. Full description of these schemes can be found in the FLUENT 6.2 User Manual (2002). A brief explanation on geometric reconstruction is explained here as it is recommended in FLUENT 6.2.

In the FLUENT 6.2 User Manual, the geometric reconstruction scheme approach standard interpolation to obtain face fluxes as the cell fills with one phase or another. Only at the cell near the interface between two phases is this method widely applied. This method is generalized for unstructured meshes and obtained from the work of Young (Piecewise linear 1982). It is assumed that for the calculation of the advection of fluid through the cell faces, the interface between two fluids has a linear slope within each cell. The following are the procedures of reconstruction on interface using geometric reconstruction scheme:

i.   First, calculation of position of the linear interface relative to the centre or each partial cell is done based on values of volume fraction.

ii. Second, the advection amount of fluid through each face is calculated using computed interface representation. Normal and tangential velocity distribution on the faces is also used as information in calculating the advection of fluid.

iii. Finally, each cell of volume fraction is calculated by computing the balance fluxes from the previous step.

Aside from the steady state problem, the geometric reconstruction scheme also available for transient cases. User defined functions can be used in order to obtain the desired solution of interest.

3.1.4 Effect of surface tension in volume of fraction

In VOF, attraction between the molecules in a fluid causes the rises of surface tension effect. This surface tension is a force only acting on the surface. The surface tension model in FLUENT is the continuum surface force (CSF) model proposed by Brackbill et al. 1992. In this model, the addition of surface tension to the VOF calculation results in a source term in the momentum equation. The pressure drop across the surface depends upon the surface tension coefficient and the surface curvature as measured by two radii in an orthogonal direction, R1 and R2:

$$\Delta p = p_2 - p_1 = \sigma \left(\frac{1}{R_1} + \frac{1}{R_2} \right) \tag{9}$$

where,

$$\kappa = \left(\frac{1}{R_1} + \frac{1}{R_2} \right), \tag{10}$$

hence,
$$\Delta p = \sigma \kappa \tag{11}$$

In equation $\Delta = p$ is the pressure drop across the surface, σ is the surface tension and κ is the mean curvature.

The formulation of CSF model is used in surface curvature computation and it is located from the local gradient in the surface normal at the interface. For example, let n be the surface normal and the gradient is α_q. Then, the volume fraction will be:

$$n = \nabla \alpha_q \tag{12}$$

And the curvature represented by divergence of the unit normal, $\hat{n}$ $\kappa = \nabla . \hat{n}$

where
$$\hat{n} = \frac{n}{|n|} \tag{13}$$

With this equation, the pressure variation across the interface is assumed linear and given by:

$$p_2 = p_1 + \sigma \kappa (\Delta \alpha_q) \tag{14}$$

The force at the surface can be expressed as a volume force using divergence theorem:

$$F_{vol} = \sum_{pairsij, i<j} \sigma_{ij} \frac{\alpha_i \rho_i \kappa_j \nabla \alpha_j + \alpha_j \rho_j \kappa_i \nabla \alpha_i}{\frac{1}{2}(\rho_i + \rho_j)} \tag{15}$$

And if only two phases are present in a cell, then $\kappa_i = -\kappa_j$ and $\nabla \alpha_i = -\nabla \alpha_j$, this equation can simplified to:

$$F_{vol} = \sigma_{ij} \frac{\rho \kappa_i \nabla \alpha_i}{\frac{1}{2}(\rho_i + \rho_j)} \tag{16}$$

ρ is the volume averaged density computed using Equation 4 . The simplified volume force shows that the surface tension source term for a cell is proportional to the average density in the cell.

3.2 Boundary conditions

The liquid was considered Newtonian with viscosity μ, density ρ and surface tension α flows in the region. The film thicknesses were obtained from the lipo lipid thickness in the human airway and it was set to be constant from the entrance of the tracheobronchial tree and uniform to its outlet. The membranes were patched with a similar thickness of the film, in order to reduce the computation period. A very fine mesh was created at the edge of the lung walls. The number of cells used was 198200 and a finer grid was placed near the thin film where the location of the interface was used. Two velocities inlets and a pressure outlet were used as boundary condition inputs. Inlet velocities were separated into two parts; small inlet and big inlet. The small inlet was selected for the thin film boundary condition and the flows maintained at 0.0001 m/s. Meanwhile the big inlet was set to 0.001 m/s to develop consistent laminar flows throughout the geometry.

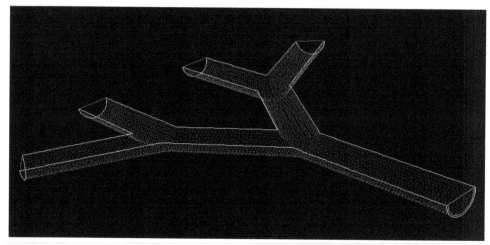

Fig. 4. Patch region in the semi symmetrical airway model

These velocities were applied as the Reynolds number for thin films considered to be ≈ 1. The density of water and viscosity of the lipo liquid was fixed to be 1023 kg/m3 and 0.00093 kg/ms and for air the density was 1.225 kg/m3 and viscosity 1.7894 x 10-5 kg/ms. The calculations were performed transiently using FLUENT 6.2 until the steady state system was reached. Initial the study was done to anticipate that the solution could be run in a steady state so as to avoid time consuming transient calculations, but for better results for the interface it was run under transient mode. A grid dependency study was done until it was found that the location of the interface was not changing with the grid refinement. The time step was set to be 0.001s and simulation was done until 6750 iteration for the solution to converge. Later the iteration was reduced to 4600 iteration as the convergence residual was achieved sooner than expected. After total time of 38 hours, all the simulations reached convergence state. In order to achieve the state of convergence, severe computational time step needed to be selected. Initially time step was set to be 0.00001s and it not only prolonged the state of convergence, but also required more computational time. An assumption was made to set the time step to be 0.001s after 2500 iterations with results not showing prominent converging state of residuals. It was important to select the correct values of velocity based on thin film liquid flow. As per the initial guesses done with different velocity magnitude, the results show the region of multiphase fluids washed by the greater velocity of air. Based on several recalculations, attempts were made to maintain the thin film velocity to be as low as 1 (Reynolds number) in order to achieve the desired profile.

4. Results and discussion

The simulations on lipo lipid with interface were done with two different selections of velocity intake based on nominal inhalation in the human body. To obtain more detailed studies on each of these cases, the effect of surface tension were also included later to the simulation to provide better understandings of the membrane function of the pulmonary surfactant (lipo lipid in the human airway). A contour plot of lipid air volume fraction is shown in Figure 5. The contours are from the calculation in which the effect of surface tension was neglected.

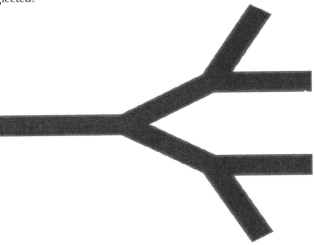

Fig. 5. Contour of volume of fraction with surface tension neglected of airway mode

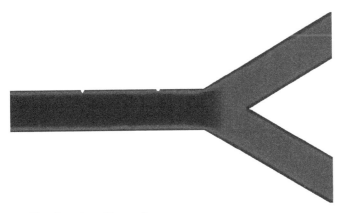

Fig. 6. Velocity profile of contour Figure 6

There are only small effects of interface changes shown for the simulation above. Near the edge of both membranes bordering adhesive to the human lung wall, the region bordering both fluids shows a minimum deformation over each fluid. Conversely there is some numerical diffusion involved where smeared effects can be seen at the interface resolution. In order to improve this factor, finer grids are able to solve this problem, but it occupies more computation time and it was not attempted here. This simulation was repeated by adding surface tension factor. The surface tension coefficient was taken to be 2 Nm as per lipo lipid configuration in the human body to obtain the effect of surface tension at the interface. The contours of volume of fraction are shown in Figure 7 based on the interface region taken approximately F=0.5 from the VOF calculation.

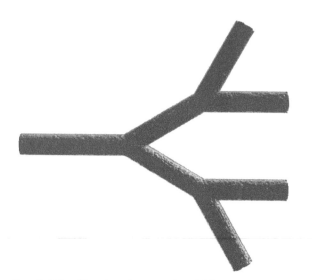

Fig. 7. Interface section F=0.5 without surface tension, 0 Nm

In order identify the effect of the pulmonary surfactant, the edge perimeter of each simulation was selected to obtain the interface section for both, with and without surface tension. Figure 8 and Figure 9 will provide details of the edge perimeter membrane interface of pulmonary surfactant. The figure was studied from the entrance of the airway up until the first bifurcation section and the pulmonary effects were uniform throughout the model. As can be seen from the figure, the effect of surface tension plays a vital role in surface reconstruction. It is very prominent to see that as the value of surface tension increases from 0 to 2 (0 = no surface tension), the effect on the interface is to deliver a smoother curve on the edge of bifurcation and wall. It can be concluded that with surface tension effect, the volume fraction creates an even structure when surface tension increases. This proves the understanding or surface tension effect where the results show smoother effects as surface tension is applied. This also agreed with human lipo lipid membrane function where the existence of the membrane as thin liquid film provides a better breathing cycle and protect from injuries by preventing the collapse or absences of surfactant.

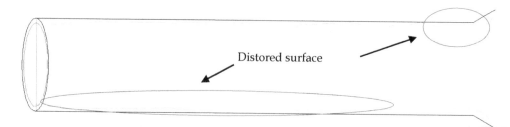

Fig. 8. Model without surface tension 0 Nm

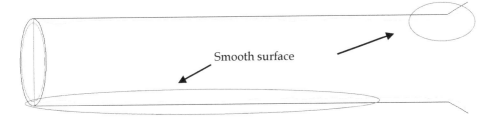

Fig. 9. Model with surface tension 2 Nm

The iso-surface obtained above was computed in a polynomial condition to obtain a better understanding between the two simulations. It is understood that the two phases are separated by an interface and the normal velocity of the interface must denote the value of zero. From the results obtained, the interface region of the volume fraction varies from 1 (lipo lipid) to zero (air). It can be concluded that the interface can be located along this region. Starting from the volume fraction contour plot, such as the one shown in Figure 10, the coordinates and velocities for each point (defined by the grid) on the assumed interface can be obtained.

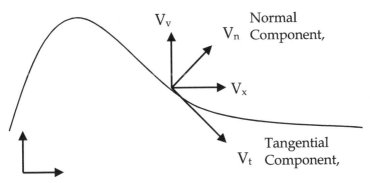

Fig. 10. Schematic of the interface

Considering a point on the interface, see Figure 10, the CFD calculated velocity components (V_x, V_y) correspond to the x, y values of the coordinate system. Normal and tangential velocities need to be calculated to evaluate the interface boundary conditions. An interface curve was fitted through sets of computed points corresponding to each of the volume fraction values investigated. The height of the interface can be expressed through the corresponding value of y along coordinate distance x as:

$$y = f(x) \tag{17}$$

This interface function of y(x) defined, which is a continuous function, can then be differentiated to obtain the interface slope:

$$\frac{dy}{dx} = f'(x) \tag{18}$$

From the geometry, values of the unit normal vector can be obtained as follows:

$$\hat{n}_x = -\frac{dy/dx}{\sqrt{1+\left(\dfrac{dy}{dx}\right)^2}} , \quad \hat{n}_y = \frac{1}{\sqrt{1+\left(\dfrac{dy}{dx}\right)^2}} \tag{19}$$

Similarly the unit tangential vectors are given by:

$$\hat{n}_y = -\frac{dy/dx}{\sqrt{1+\left(\dfrac{dy}{dx}\right)^2}} , \quad \hat{n}_x = \frac{1}{\sqrt{1+\left(\dfrac{dy}{dx}\right)^2}} \tag{20}$$

Having obtained these unit vectors, the normal and tangential velocity components of the interface are respectively given by:

$$V_{\hat{n}} = V_x\,\hat{n}_x + V_y\,\hat{n}_y \tag{21}$$

$$V_{\hat{t}} = V_x\,\hat{t}_x + V_y\,\hat{n}_y \tag{22}$$

The calculations of first derivative dy/dx ensure the accuracy of the result. Here dy/dx represent the slope at any point x on the curve. Consequently, it is important to determine the precise interface curve representation $f(x)$ that passes through all data points.

Computation of normal velocity at the interface was done based on equation 21 and was repeated for both cases of volume of fraction on the conducted simulation. The objective of the investigation was to verify the accurateness of the obtained interface and its sensitivity to selection of significant volume fraction. The value of 0.5 of volume of fraction used here was based on previous work that shows at this particular value the interface can be presented more significantly. Using equation 21, the normal velocity along the x direction of the interface was calculated. The plot of normal velocity with respect to x direction is shown in Figure 11.

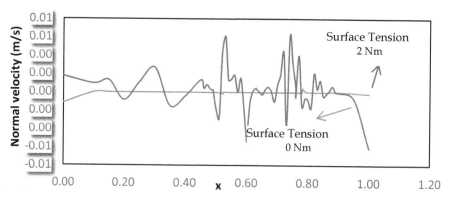

Fig. 11. Graph of normal velocity of iso -0.5 of both simulations

The comparison was done between the plot of normal velocity without surface tension and with surface tension. From Figure 11 the surface tension creates a smoother surface. The values of normal velocities of each case provide sufficient value of zero. It is understood that the interface region is zero normal velocity and with the application of the surface tension in the simulation it denotes as the numerical value. The results obtained clearly indicate that using VOF in FLUENT facilitate a good approach in determining the interface of new surface reconstruction. In the cases of using VOF method in the thin liquid film flow, this provides significant results on pulmonary surfactant condition in thin film flow.

5. Conclusion

In this chapter, a 3D model of the human tracheobronchial tree was investigated by using commercial software FLUENT 6.2. The major study was done to understand the behaviour of pulmonary surfactant in the human airway whereby the surface tension presences in airways are significant and without the existence of surfactant, it will severely affect the condition of inhalation. Apart from this, presenting the effects of surface tension in VOF provide significant results with application in thin liquid film studies. Finally, the use of the VOF model to calculate thin film was considered. It is a known fact that in a thin film flow, surface tension plays a vital role and is prominent in controlling the film surface. The results of VOF show that application of this method in the human airway with thin film flow accommodates a high level of accuracy. The conclusion was made that use of VOF in thin film flow is appropriate in identifying the real interface using volume fraction of VOF and the pulmonary surfactant in the human airway plays a major role for inhalation.

6. Acknowledgment

I have taken great effort in writing this book chapter, however, it would not have been possible without the kind support and help of many individuals and my organization. I would like to extend my sincere thanks to all of them. I am highly indebted to Bumi Armada Berhad and my family members for their guidance and constant motivation as well as for providing necessary support in completing this chapter. I would like to express my special gratitude and thanks to the publisher and the University of Rijeka, Croatia, for giving me such attention and time.

7. References

Allen G.M, Shortall B.P, Gemci T, Corcoran T.E, and Chigier N.A, (2004). Computational simulations of airflow in an in vitro model of the pediatric upper airways, *ASME Journal of Biomechanical Engineering* 126, pp 604–613

Brackbill, J.U., Kothe, D.B. and Zemach, C. (1992). A continuum method for modelling surface tension, *Journal of Computational Physics* 100, pp 335–354

Caro C.G, Schroter R.C, Watkins N, Sherwin S.J, and Sauret V, (2002). Steady inspiratory flow in planar and non planar models of human bronchial airways, *Proceedings of the Royal Society*, pp 458:791 - 809.

Cerne, G., Petelin, S. and Tiselj, I. (2000). Upgrade of the VOF method for the simulation of the dispersed flow, *Proceedings of the ASME 2000, Fluid Engineering Division Summer Meeting* Boston, Massachusetts

Farkas A, and Balashazy I, (2007). Simulation of the effect of local obstructions and blockage on airflow and aerosol deposition in central human airways, *Journal of Aerosol Science* 38: pp 865–884

FLUENT 6.2 User Manual, (2002)

Gemci T, Ponyavin V, Chen Y, Chen H, Collins R, (2008). Computational model of airflow in upper 17 generations of human respiratory tract, *Journal of Biomechanics*, Volume 41, Issue 9, pp 2047-2054

Martonen, T. B., Fleming, J., Schroeter, J., Conway, J., & Hwang, D. (2003). In silico modelling of asthma, *Advanced Drug Delivery Reviews*, *55*, pp 829–849

Schmidt A, Zidowitz S, Kritete A, Denhard T, Krass S, and Peitgen H.O, (2004). A digital reference model of the human bronchial tree, *Computerized Medical Imaging and Graphics*, pp 28:203-211

Versteeg H. K., and Malalasekera W. Book - An introduction to computational fluid dynamics, Pearson: (1995)

Youngs, D.L. (1982). Time-dependent multi-material flow with large fluid distortion, in Numerical Methods for Fluid Dynamics, *K.W.Morton and M.J.Baines (Eds.)*, pp 273-285

2

3D Particle Simulations of Deformation of Red Blood Cells in Micro-Capillary Vessel

Katsuya Nagayama[1] and Keisuke Honda[2]
Kyushu Institute of Technology
Hitachi Cooperation
Japan

1. Introduction

With the increase in arteriosclerosis, thrombosis, etc., in order to find out the cause, research of the flow characteristic of blood attracts attention. As for the analysis of the flow phenomenon of the RBC (Red Blood Cell or Erythrocyte), the numerical simulation (Wada et al., 2000, Tanaka et al., 2004) as well as experiment observation (Gaehtgens et al., 1980) is becoming a strong tool. Particle methods, such as SPH method (Monaghan J., 1992) and the MPS method (Koshizuka, 1997), treats both solid and liquid as particles, and can be applied to complicated flow analysis. When applying a particle method to the flow analysis of RBC, RBC is divided into the elastic film and internal liquid, and its deformation was analyzed in detail (Tanaka et al., 2004, Tsubota et al., 2006).

The RBC which is actually flowing in our body occupies 40-60% by volume ratio of blood (hematocrit), and is numerous. The objective of our research is clarifying the flow characteristic of the blood flow containing many RBCs. We reported preliminarily simulation of 2D blood flow (Nagayama et al., 2004), where many RBCs were simply treated as a lump of an elastic particle, the flow was analyzed qualitatively. Moreover, three dimensional RBC was modelled with double structure, and the RBC shape in flow was more realistic (Nagayama et al., 2005). The relation of the blood vessel diameter and the blood-flow with many RBCs was studied by 2D model (Nagayama, 2006) and by 3D model (Nagayama et al., 2008a). The model was also applied to 3D blood flow in capillary bend tube (Nagayama et al., 2008b).

The objective is to understand the fundamental flow phenomenon in a blood vessel. In this paper, 3 dimensional blood flows with RBCs in capillary tube were simulated.

In Section 2, simulation model was described. And the shape of single red blood cell in static fluid was shown.

In Section 3, blood flows with RBCs in capillary straight tube were simulated. And the relations of the blood vessel diameter and the hematocrit were investigated. Furthermore, transient phenomena of interacting red blood cells and their shape were investigated.

In Section 4, the model is applied to the capillary vessel flow at finger tip edge. The capillary vessel is modelled as two cases. One case is bent tube and another is bent and twisted tube, and RBC deformation were investigated.

2. Model descriptions

The particle model used for simulation is described. Then calculation conditions will be explained.

2.1 Mathematical descriptions

Particle method considers the interaction between particles and pursues motions of particles in Lagrangian way. Instead of NS equation, a momentum equation in particle model (1) consists of inertial force, inter-particle force, viscous diffusion and external force. Inter-particle force is attracting force or repulsing force between particles using particle pressure as shown in equation (2), so that to keep the density uniform in the domain.

Spring model is also considered for the elastic RBC surface. For the viscous diffusion term, MPS method (Koshizuka, 1997) is used. As for the external force, pressure difference between both ends of blood vessel was taken into consideration. The symbols are, u: velocity vector, t: time, ω: weighting function, n: number density, r: position, d and λ: constants. In addition i: particle number, j: surrounding particle number, 0: basic condition. RBC film particle is tied by surrounding particles using springs. In addition, resistance against bending is modelled as the force to the center of mass of surrounding particles. In addition, damping force is treated as viscous force in Eq. 1. The size of RBC is about 8 µm in diameter and 3 □µm in thickness.

$$\frac{\partial \mathbf{u}_i}{\partial t} = -\frac{1}{\rho}\frac{d}{n^0}\sum_{j\neq i}\left[\left(P_j - P_i\right)\omega_{ij}\frac{\mathbf{r}_{ij}}{\left|\mathbf{r}_{ij}\right|^2}\right] + v\frac{2d}{n^0\lambda}\sum_{j\neq i}\left[\omega_{ij}\left(\mathbf{u}_j - \mathbf{u}_i\right)\right] + \mathbf{F} \tag{1}$$

$$P_i = \frac{1}{\kappa}\left(1 - \frac{n_0}{n_i}\right) \tag{2}$$

2.2 RBC shape in static fluid

RBC is double structure: surface film and plasma liquid inside. RBC film particle is tied by surrounding film particles by springs with coefficient of 7.52×10^{-2}N/m as shown in Fig. 1. In addition, resistance against bending with coefficient of 3.76×10^{-4}N/m, is modelled as the force to the center of mass of surrounding particles. Surface area is 140µm² and volume is 90µm³. In the simulation, starting from sphere shape and removing 42% of plasma particle inside, the shape of RBC is formed. Fig.2 is the simulated RBC shape in static fluid. The size of RBC is about 8 µm in diameter and 3 µm in thickness. RBC shape will change with flow in blood vessel.

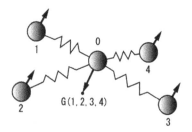

Fig. 1. Elastic film model

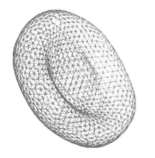

Fig. 2. Simulated RBC shape in static fluid

3. Deformation of RBCs in various inner diameter capillaries and hematocrit

Deformation of RBCs in various inner diameter capillaries and hematocrit was studied. Next, transition from single-file to multi-file flow as a function of hematocrit in capillaries of various diameters was discussed.

3.1 Simulation conditions

Simulation conditions are shown in Table 1. Simulations were carried out using normalized value. The velocity is normalized by 1 mm/s and the length is normalized using 10 μm. Physical properties are also shown in Table 1. Total simulation time is 0.3 s (300000 iterations with time step1 μs), which is enough to reach stable flow.

Cases for simulation with various inner diameter (ID) capillaries and hematocrit are shown in Table 2. To study transition phenomena from single-file to multi-file flow, hematocrit 0.24-0.54 and capillaries of various inner diameters 5.5-8.7 μm were chosen for simulation.

Velocity of normalization	1	[mm/s]
Length of normalization	10	[μm]
Viscosity	0.001	[Pa s]
Density	1000	[kg m3]
Simulation time	0.3	[s]
Elasity of stretching	7.52 E-04	[N/m]
Elasity for bending	2.63 E-05	[N/m]

Table 1. Simulation conditions

Case	ID [μm]	Ht
(a)	5.5	0.31
(b)	7.37	0.24
(c)	7.37	0.49
(d)	8.5	0.21
(e)	8.7	0.54

Table 2. Cases for simulation

3.2 Results for cases in various inner diameter capillaries and hematocrit

In this section, first of all, results of deformation of RBCs in various ID and hematocrit will be shown. Next, transition from single-file to multi-file flow will be discussed. The deviation of RBC distribution in a capillary blood vessel will also be shown.

3.2.1 Deformation of RBCs in various inner diameter capillaries and hematocrit

Simulation conditions are shown in Table1. Simulations were carried out for 5 cases with various inner diameter capillaries and hematocrit. Results are shown in Fig. 3 and the RBC shape was studied for each cases.

In case of (a) ID=5.5 μm Ht=0.31 (narrowest capillary), RBC flows in lines (single-file flow). RBC contacts with wall and deforms to consistently non-axisymmetric rocket shape 'torpedo' exhibiting a membrane-fold which extends from the open rear-end along one side toward the leading end (Gaehtgens et al., 1980).

In case of (b) ID=8.5 μm Ht =0.2, RBC flows in lines (single-file flow), rarely contact with the wall. RBCs flow at center of the blood vessel, parachute type deformation appeared.

In case of (c) ID=7.37 μm Ht =024, RBC flows basically in lines (single-file flow). RBCs do not flow at center of blood vessel. RBC interact each other, and sometimes contact with another RBC.

In case of (d) ID=7.37 μm Ht=0.49, RBC interacts (multi-file flow) with each other and contact with the wall, forming zipper shape.

In case of (e) ID=8.7 μm Ht=0.54, RBC interacts (multi-file flow) strongly with each other and contact with the wall, forming strong and complex deformation.

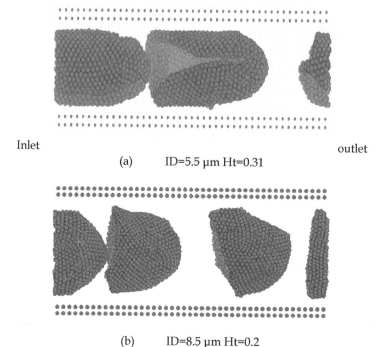

Inlet   outlet

(a)   ID=5.5 μm Ht=0.31

(b)   ID=8.5 μm Ht=0.2

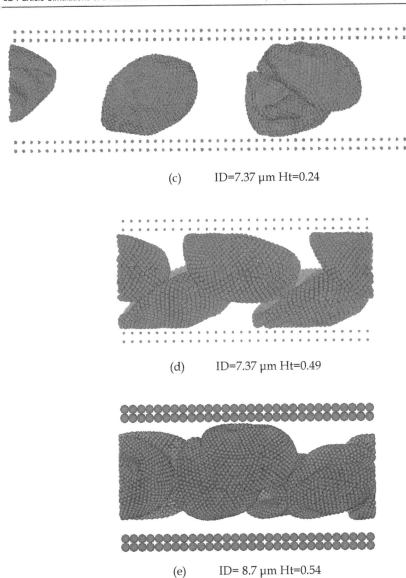

(c)   ID=7.37 μm Ht=0.24

(d)   ID=7.37 μm Ht=0.49

(e)   ID= 8.7 μm Ht=0.54

Fig. 3. Deformations of RBCs in various inner diameter capillaries and hematocrit

3.2.2 Transition from single-file to multi-file flow

Transition from single-file to multi-file flow as a function of hematocrit in capillaries of various diameters is shown in Fig.4. Overall tendency in the experiment (Gaehtgens et al., 1980) and simulation are similar qualitatively. RBCs are single-file in narrow tube and at low hematocrit, while they are multi-file as the tube diameter increases or hematocrit increases. A line in Fig.2 is Ht = 2.8/ID. By a rough classification, RBCs are single-file when Ht < 2.8/ID, while they are multi-file when Ht >2.8/ID.

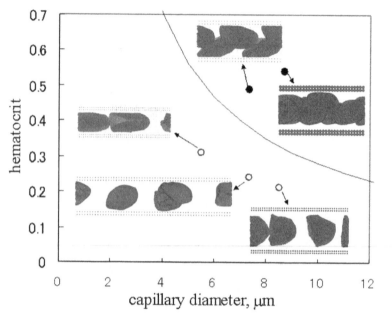

Fig. 4. Transition from single-file to multi-file flow as a function of hematocrit in capillaries of various diameters

3.2.3 Particle simulation about the deviation of RBC distribution in a capillary blood vessel

In Case (d), although RBCs were placed in line initially, they interacted and flows like zipper shape finally. This transition will be described in detail here. In Fig.5, RBC shape are shown for Case (d) ID=7.37 μm Ht =0.49, at 0 ms, 40 ms, 80 ms, 140 ms, 200 ms and 260 ms.

At 40 ms, RBCs are at the center of the blood vessel, parachute type shape, and flows in line. The back of the erythrocyte is dented, and it can also be said the bowl type.

In addition, plasma flow without the RBCs was calculated, the flow was Hagen-Poiseuille flow and velocity distribution at the cross section in pipe was parabola-shaped. When there were RBCs, RBC particles flows together due to the elastic film, and the velocity distribution was near in a trapezoid.

In 80 ms, the inclination occurs to arrangement of RBCs, and intervention happened. Uniformity in the axis center of the parachute-shaped collapsed, and intervention with a face of wall and the surrounding erythrocyte happened.

In time 140 ms, more mutual intervention of an erythrocyte was seen, and the shape is being changed complicatedly while having contact and rallying. A back RBC enters into the indent of the previous RBC, and the transfer state to the zipper type was seen.

Moreover its tendency was strengthened in time 200 ms. Back erythrocyte was entering into the indent of the previous erythrocyte, and 4 erythrocytes have ranged. At 260 ms, RBCs flow in the zipper shape, and it became stable.

As shown, RBCs flow with intervention of a tube wall and between the erythrocytes mutually. Initially RBCs flows with the parachute shape at early stage, but they begun to fluctuate and became unstable state. Mutual intervention of an erythrocyte was seen, and

the shape is being changed complicatedly while having contact and rallying. Finally RBCs flow in the zipper shape, and it became stable.
It is thought that the placement of the red blood cell changes when the condition changes. It is expected that the stable state changes by pipe diameter, a red count, the properties of matter of the red blood cell.

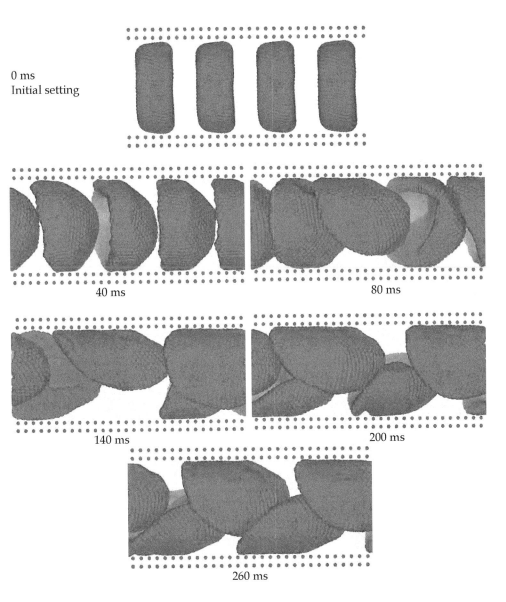

Fig. 5. RBC shape change for ID=7.37 μm Ht =0.49

4. Particle simulations of blood flow in bent and twisted capillary vessel with red blood cells

In this section, particle model is applied to simulate the capillary vessel flow in the turning point of blood vessel at finger tip. The capillary vessel is modelled as two cases: one is bent tube and another is bent and twisted tube.

4.1 Simulation conditions

Using microscope, capillary blood vessels can be observed at the finger tip as shown in Fig. 6. They change their shape depending on the health. To supply nutrition and to remove wastes, bent shape is usual and in good health. Winding and twisted shape tend to be observed in unhealthy condition such as high viscosity, although the reason is not clear. Blood flow analysis is important to study blood circulation. Here, particle model is applied to the capillary vessel flow to clarify the flow characteristics and mechanism of shape change.

Physical properties of plasma are density 1030 kg/m³, and viscosity 1.2 mPa·s. Pressure gradient of 100 kPa/m is applied which cause blood velocity about 1 mm/s. (Note pressure gradient is fixed and the velocity depends on the tube shape and RBC number). Reynolds number is 0.86×10^{-2} and flow field is laminar. Periodic boundary condition is used at inlet and outlet, exit particles are supplied from inlet again.

Two type of basic shape of the capillary vessel is modelled as shown in Fig.7. One is bent tube and another is bent and twisted tube. All the RBC, blood vessel and plasma fluid are modelled as particles and Fig.2 is the initial setting of particles (plasma particles are not shown). Bent tube is inner diameter 6.82μm, length of straight portion 37.2μm. Twisted tube is inner diameter 7.37μm, height 90μm.

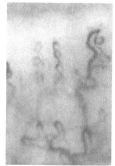

Microscope   Bent   Winding and twisted

Fig. 6. Observed blood vessel at the finger tip

Particles are set every 0.62 μm in bent tube and 0.67 μm in twisted tube. At the surface of RBC, particle density is high to increase the accuracy of RBC shape. Cases for simulation with various RBC number for bent tube and twisted tube are shown in Table 3. To study the effect of RBC on flow field, cases with different number of RBC are simulated. Volume ratio (Hematocrit) of one RBC is 2.23% in bent tube and 1.57% in twisted tube. Hematocrit is 31% in bent tube and 38 % in twist tube, which is close to the typical range of 40-50%. Total particle number including plasma, wall and RBC is minimum 29490 in plasma flow in bent tube, maximum 85304 in 24 RBC flow in twisted tube. Total simulation time is 100 ms (100000 iterations with time step1 s).

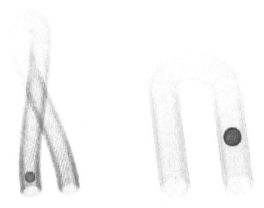

Fig. 7. Model for bent tube and twisted tube (with 1RBC)

Tube type	Number of RBC	Particle number for simulation
Bent tube	0	29490
	1	30742
	14	46995
Twisted tube	0	54446
	1	55728
	24	85304

Table 3. Cases for simulation

4.2 Results for blood flow in bent and twisted capillary vessel

In this section, results of RBC deformation for two cases: one is bent tube and another is bent and twisted tube will be shown for one RBC and many RBCs.

4.2.1 RBC deformation in bent tube

Simulated RBC shape in bent tube is shown in Fig.8. Fig.8 (a) is the result of one RBC case, and RBC shape is shown every 5 ms up to 40 ms. At 5 ms in the straight portion, RBC is parachute shape due to the fast flow at the tube center and slow flow close to the wall, which is typical in capillary tube. Velocity field around RBC is trapezoid, while parabolic Poiseuille flow in plasma flow far from RBC. At 10 ms to 20 ms, RBC is passing through the bent portion, starts to deform to asymmetric parachute shape. RBC particle tends to pass quickly inside the bent and slowly outside the bent, due to the flow length difference. After 25 ms, RBS shape tends to recover to symmetric parachute shape.

Fig.8 (b) is the snapshot of 14 RBC case at t=80 ms. RBCs tend to flow at the tube center, and the shape is between rocket and parachute. RBCs interact each other and tend to go in the back end of another RBC. When passing through the bent, RBC tends to keep the dent inside the bent.

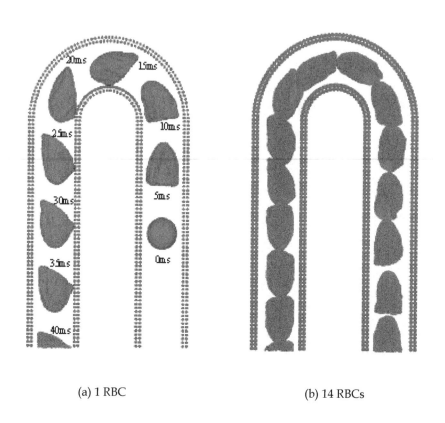

(a) 1 RBC   (b) 14 RBCs

Fig. 8. Simulated RBC shape in bent tube

4.2.2 RBC deformation in twisted tube

Simulated RBC shape in twisted tube is shown in Fig.9 Fig.9 (a) is the result of one RBC case, and RBC shape is shown at 20, 40, 50, 60, 80 ms. RBC is parachute shape slightly asymmetric before the bent at 20 ms. At 40 ms to 60 ms, RBC is passing through the bent portion and deforms to asymmetric shape (between flat and parachute shape) due to the twist and bend flow. RBC particle tends to pass quickly inside the bend and slowly outside the bend, due to the flow length difference. After 60 ms, RBS shape tends to recover to symmetric parachute shape.

Fig.9 (b) is the snapshot of 24 RBC case at t=90 ms. RBC shapes are quite uneven by the strong interaction due to the twist and bent.

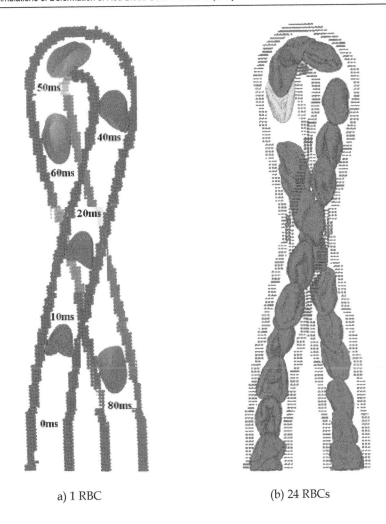

a) 1 RBC   (b) 24 RBCs

Fig. 9. Simulated RBC shape in twisted tube

5. Conclusion

3 dimensional particle model is applied to the capillary straight tube flow.

1. Deformations of RBCs in various ID (inner diameter) capillaries and hematocrit were studied. In case of ID=5.5µm Ht=0.31, RBC flows in lines contacting with the wall and deforms to consistently non-axisymmetric rocket shape. In case of ID=8.5µm Ht =0.2, RBCs flow at center of the blood vessel, parachute type deformation appeared. In case of ID= 7.37µm Ht = 0.49, RBC interacts (multi-file flow) with each other and contact with the wall, forming zipper shape. In case of ID=8.7µm Ht =0.54, RBC interacts (multi-file flow) strongly with each other and contact with the wall, forming strong and complex deformation.

2. Transition from single-file to multi-file flow as a function of hematocrit in capillaries of various diameters is studied. RBCs are single-file in narrow tube and at low hematocrit, while they are multi-file as the tube diameter increases or hematocrit increases.

3. RBC shape change in time was studied for a case of ID= 7.37μm Ht = 0.49 in details. At first, RBCs were flowing in line like a parachute. But after that the shape fluctuated gradually, and it became stable in this case in the zipper state finally. A stable state is expected that it also changes by the pipe diameter, the number of erythrocytes and the physical properties of the RBCs.

4. Particle simulations were applied to the capillary vessel flow at finger tip. In case of bent tube, the RBCs is initially parachute shape at straight tube and then deforms to asymmetric parachute shape. In case of bent and twisted tube, initially RBC is parachute shape and then deforms to asymmetric shape (between flat and parachute shape) due to the twist and bend flow.

6. References

Wada, Kobayashi, Takahashi and Karino (2000). A numerical simulation of the deformation of an erythrocyte, *Japanese Mechanical Engineering Congress*, MECJ-05, No.1226 pp.287-288, Japan, 2000

Tanaka N., Takano T. and Masuzawa T. (2004). 3-dimensional micro-simulation of blood flow with SPH method, *Japanese Fluid Engineering Conference*, JSME, No. 712, Japan, 2004

Gaehtgens P., Duhrssen C. and Albrecht KH. (1980). Motions, Deformation and Interaction of Blood Cells and Plasma During Flow Through Narrow Capillary Tubes, *Blood cells* Vol.6, 799-812, 1980

Monaghan J. (1992). *Annu. Rev. Astrophys.*, No.30, pp.543-574, 1992

Koshizuka S., *Computational Fluid Dynamics* (1997). in Japanese, Baihuukan, Japan, 1997

Tsubota K., Wada S. and Yamaguchi T. (2006). Particle method for computer simulation of red blood cell motion in blood flow, *Computer Methods and Programs in Biomedicine*, Vol. 83, pp. 139-146, 2006.

Nagayama K. and Tanaka K. (2004). Particle Simulations of Two Phase Blood Flow with Red Blood Cell, *Japanese Fluid Engineering Conference*, JSME, No.G808, Japan, 2004, ISSN 1348-2882

Nagayama K. and Tanaka K. (2005). Particle Simulations of three dimensional blood flow with a blood cell, *Proceedings of 2005 Annual Meeting*, Japan Society of Fluid Mechanics AM05-17-007, Japan, 2005

Nagayama K. (2006), Particle Simulations of Blood flow in Vein with Many RBCs, *International Proceedings by Medimond from World Congress of Biomechanics*, pp. 557-562 Munich, Germany, 2006, Volume ISBN 88-7587-270-8, CD ISBN 88-7587-271-6

Nagayama K. and Honda K. (2008a), Particle Simulations of the deformation of red blood cells in a capillary vessel. *Proceedings of the 12th Asian Congress of Fluid Mechanics*, Daejeon, Korea, 18-21 August 2008

Nagayama K. and Honda K. (2008b), Particle Simulations of Blood Flow in Bent and Twisted Capillary Vessel with Red Blood Cells, *Proceedings of the TFEC*, Sapporo, Japan, 14-16 September 2008

Modelling Propelling Force in Swimming Using Numerical Simulations

Daniel A. Marinho[1,2], Tiago M. Barbosa[2,3], Vishveshwar R. Mantha[2,4],
Abel I. Rouboa[2,5] and António J. Silva[2,4]

[1]*University of Beira Interior, Department of Sport Sciences, Covilhã*
[2]*Research Centre in Sports, Health and Human Development, Vila Real*
[3]*Polytechnic Institute of Bragança, Department of Sport Sciences, Bragança*
[4]*University of Trás-os-Montes and Alto Douro, Department of Sport Sciences,
Exercise and Health, Vila Real*
[5]*University of Trás-os-Montes and Alto Douro,
Department of Engineering, Vila Real*
Portugal

1. Introduction

In the sports field, numerical simulation techniques have been shown to provide useful information about performance and to play an important role as a complementary tool to physical experiments. Indeed, this methodology has produced significant improvements in equipment design and technique prescription in different sports (Kellar et al., 1999; Pallis et al., 2000; Dabnichki & Avital, 2006). In swimming, this methodology has been applied in order to better understand swimming performance. Thus, the numerical techniques have been addressed to study the propulsive forces generated by the propelling segments (Rouboa et al., 2006; Marinho et al., 2009a) and the hydrodynamic drag forces resisting forward motion (Silva et al., 2008; Marinho et al., 2009b).

Although the swimmer's performance is dependent on both drag and propulsive forces, within this chapter the focus is only on the analysis of the propulsive forces. Hence, this chapter covers topics in swimming propelling force analysis from a numerical simulation technique perspective. This perspective means emphasis on the fluid mechanics and computational fluid dynamics methodology applied in swimming investigations. One of the main aims for performance (velocity) enhancement of swimming is to maximize propelling forces whilst not increasing drag forces resisting forward motion, for a given trust. This chapter will concentrate on numerical simulation results, considering the scientific simulation point-of-view, for this practical application in swimming.

Basically, numerical simulations consist of a mathematical model that replaces the Navier-Stokes equations with discretized algebraic expressions that can be solved by iterative computerized calculations. The Navier–Stokes equations describe the motion of viscous non-compressible fluid substances. These equations arise from applying Newton's second law to fluid motion, together with the assumption that the fluid stress is the sum of a diffusing viscous term (proportional to the gradient of velocity), plus a pressure term. A solution of the Navier–Stokes equations is called a velocity field or flow field, which is a description of the velocity of the fluid at a given point in space and time. Numerical simulations are based on the finite volume approach, where the equations are integrated over each control volume. It is required to discretize the spatial domain into small cells to form a volume mesh or grid, and then apply a suitable algorithm to solve the equations of motion (Marinho et al., 2010a). Additionally, several studies reported the importance of numerical simulations on testing and experimentation, reducing the total effort required in the experimental design and data acquisition (Lyttle & Keys, 2006; Bixler et al., 2007). For instance, Lyttle and Keys (2006) referred that these numerical simulations can provide answers into many complex problems that have been unobtainable using physical testing techniques. One of its major benefits is the possibility to test many variations until one arrives at an optimal result, without physical/experimental testing.

Although some doubts on the accuracy of numerical simulations results, this numerical tool has been validated as being feasible in modelling complicated biological fluid dynamics, through a series of stepwise baseline benchmark tests and applications for realistic modelling of different scopes for hydro and aerodynamics of locomotion (Liu, 2002). Bixler et al. (2007) studied the accuracy of numerical analysis of the passive drag of a male swimmer. Comparisons of total drag force were performed between a real swimmer, a digital model of this same swimmer and a real mannequin based on the digital model. Bixler et al. (2007) found drag forces determined from the digital model using numerical simulations to be within 4% of the values assessed experimentally for the mannequin, although the mannequin drag was found to be 18% less than the real swimmer drag. In fact, this study has reinforced the idea of the validity and accuracy of numerical simulations in swimming research. Some differences were due to little body movements during the gliding position and to differences between the swimmer and the model since the swimmer's skin is flexible while the mannequin's skin is rigid. So, it is usually assumed that numerical simulations have ecological validity even for swimming research.

In the first part of the chapter, we introduce the issue, the main aims of the chapter and a brief explanation of the computational fluid dynamics methodology. Then, the contribution of different studies for swimming using numerical studies and some practical applications of this methodology are presented. During the chapter the authors will attempt to present the computational fluid dynamics data and to address some practical concerns to swimmers and coaches, comparing as well the numerical data with other experimental data available in the literature.

2. Numerical simulation methodology

2.1 Mathematical formulation

The flow around the swimmer seems to be turbulent (Sanders, 2001; Toussaint & Truijens, 2005). Due to that reason Reynolds averaged Navier–Stokes equations with the Boussinesq hypothesis to model the Reynolds stresses are usually used (Hinze, 1975). The closure problem of the turbulent modeling is solved using k–epsilon model with appropriate wall functions. The system of equations for solving three-dimensional, incompressible fluid flow in steady-state regime is as follows:

$$div V = 0$$

(1)

$$\frac{\partial V}{\partial t} \pm V.\nabla V + \nabla p \pm \nabla \left(v + c_\mu \frac{k^2}{\varepsilon} \right)\left(\nabla V + \nabla V^t \right) = 0 \qquad (2)$$

$$\frac{\partial(\rho k)}{\partial t} + \frac{\partial(\rho V_x k)}{\partial x} + \frac{\partial(\rho V_y k)}{\partial y} + \frac{\partial(\rho V_z k)}{\partial z} = \frac{\partial \left(\frac{\mu_t}{\sigma_k} \frac{\partial k}{\partial x} \right)}{\partial x} + \frac{\partial \left(\frac{\mu_t}{\sigma_k} \frac{\partial k}{\partial y} \right)}{\partial y} + \frac{\partial \left(\frac{\mu_t}{\sigma_k} \frac{\partial k}{\partial z} \right)}{\partial z} + \mu_t \Phi - \rho \varepsilon \qquad (3)$$

$$\frac{\partial(\rho k)}{\partial t} + \frac{\partial(\rho V_x \varepsilon)}{\partial x} + \frac{\partial(\rho V_y \varepsilon)}{\partial y} + \frac{\partial(\rho V_z \varepsilon)}{\partial z} = \frac{\partial \left(\frac{\mu_t}{\sigma_\varepsilon} \frac{\partial \varepsilon}{\partial x} \right)}{\partial x} + \frac{\partial \left(\frac{\mu_t}{\sigma_\varepsilon} \frac{\partial \varepsilon}{\partial y} \right)}{\partial y} + \frac{\partial \left(\frac{\mu_t}{\sigma_\varepsilon} \frac{\partial \varepsilon}{\partial z} \right)}{\partial z} + \mu_t \frac{\varepsilon}{k} \Phi - C_2 \frac{\rho \varepsilon^2}{k} \qquad (4)$$

Where k is the turbulent kinetic energy and ε is the turbulent kinetic energy dissipation ratio. V_x, V_y and V_z represent the x, y and z components of the velocity V. μ_t is the turbulent viscosity and ρ represents the fluid density. v is the kinematic viscosity, Φ is the pressure strain, C_2, C_μ, σ_ε and σ_k are model constants, 1.92, 0.09, 1.30 and 1.00, respectively.
The detailed terms of the k–epsilon model transport equations used during simulations are provided in user manual of Fluent documentation (Fluent, 2006).

2.2 Digital model
One of the main tasks to carryout numerical simulations is to define the digital model of the swimmer. The majority of the studies performed on this field used segments of the human body, representative of propelling segments, i.e., the hand and the forearm (Bixler & Schloder, 1996; Rouboa et al., 2006).
In order to create the three-dimensional digital model computer tomography scans of a hand and forearm segments of an Olympic swimmer were applied. With these data we converted the values into a format that can be read in *Gambit*, Fluent® pre-processor. Fluent® software is used to simulate the fluid flow around structures, allowing the analysis of values of pressure and speed around (i.e. the hand and forearm of a swimmer). With these values we can calculate force components through integration of pressures on the hand/forearm surfaces, using a realistic model of these human segments.
Eighteen cross-sectional scans of the right arm (hand and forearm) were obtained using a Toshiba® Aquilion 4 computer tomography scanner. Computer tomography scans were obtained with configuration of V2.04 ER001. A 2 mm slice thickness with a space of 1 mm was used. The subject was an Olympic level swimmer, who participated in the 2004 Olympic Games in Athens. The subject was lying with his right arm extended upwards and fully pronated. The thumb was adducted and the wrist was in a neutral position (Marinho et al., 2010b). The appropriate ethical committee of the institution in which it was performed has approved this protocol, and the subject consented to participate in this work.
The transformation of values from the computer tomography scans into nodal coordinates in an appropriate coordinate system demands the use of image processing techniques. The image-processing program used in this study was the Anatomics Pro®. This program allowed obtaining the boundaries of the human segments, creating a three-dimensional reconstruction of the swimmer hand and forearm.
At first, before processing and converting procedures the data was prepared, namely by observing the computer tomography data and erasing the non-relevant parts of the anatomical model. For example, surfaces supporting the subject were also scanned, reason

why it had to be defined the relevant points and deleted the irrelevant ones. This step was also conducted using the software FreeForm Sensable®. Finally, the data was converted into an IGES format (*.igs), that could be read by Gambit/Fluent® to define the finite volume approach through the three-dimensional surfaces (Marinho et al., 2010b).

2.3 Simulations

The dynamic fluid forces produced by the hand, lift (L) and drag (D), were measured in this study. These forces are functions of the fluid velocity and they were measured by the application of the equations 5 and 6, respectively:

$$D = C_D \frac{1}{2} \rho A v^2 \tag{5}$$

$$L = C_L \frac{1}{2} \rho A v^2 \tag{6}$$

In equations 5 and 6, v is the fluid velocity, C_D and C_L are the drag and lift coefficients, respectively, ρ is the fluid density and A is the projection area of the model for different angles of attack used in this study.

The whole domain was meshed with a hybrid mesh composed of prisms and pyramids. Significant efforts were conducted to ensure that the model would provide accurate results by decreasing the grid node separation in areas of high velocity and pressure gradients.

Angles of attack of hand models of 0°, 15°, 30°, 45°, 60°, 75° and 90°, with a sweep back angle of 0° (thumb as the leading edge) were used for the calculations (Schleihauf, 1979).

Steady-state analyses were performed using the Fluent® code and the drag and lift coefficients were calculated for a flow velocity of 2.0 m/s (Lauder et al., 2001; Rouboa et al., 2006).

We used the segregated solver with the standard k-epsilon turbulence model because this turbulence model was shown to be accurate with measured values in a previous research (Moreira et al., 2006).

All numerical computational schemes were second-order, which provides a more accurate solution than first-order schemes. We used a turbulence intensity of 1.0% and a turbulence scale of 0.10 m. The water temperature was 28° C with a density of 998.2 kg·m⁻³ and a viscosity of 0.001 kg/m/s. Incompressible flow was assumed. The measured forces on the hand models were decomposed into drag (C_D) and lift (C_L) coefficients, using equations 5 and 6.

3. Propelling force

3.1 Definition

Swimming is characterized by the intermittent application of a propulsive force (thrust) to overcome a velocity-dependent water resistance (hydrodynamic drag). The thrust is generated by a combination of arm, leg and body movements and lead to variations of thrust and velocity. Different fluctuations in thrust, drag and velocity among different techniques and different level of skills contribute to the highly variable performance in swimming. Swimming performance can be studied by analysing the interaction of propelling and resistive forces. In this sense, a swimmer will only enhance performance by minimizing resistive forces that act on the swimming body at a given velocity and/or by increasing the propulsive forces produced by the propelling segments. Furthermore, a third performance-enhancing factor would be to do this with a minimal enhancement of physiological or energetic costs (Barbosa et al., 2010).

Effective propelling force can be defined as the component of the total propulsive force acting in the direction of moving. This force is produced due to the interaction of the swimmer with the water allowing overcoming hydrodynamic drag forces resisting forward motion. Thus, it is a hydrodynamic force with the same direction of the movement but opposite to drag force.

There are several mechanisms responsible to produce propelling forces, although some of them seemed to be more efficient than others. This fact seems to distinguish swimmers of different level, since some mechanisms allow producing the same work with less energy consumption (Barbosa et al., 2010). Knowing the most effective way to produce propulsive force should play a major role in swimming training.

3.2 Relative contribution of drag and lift forces

The relative contribution of drag and lift forces to overall propulsion is one of the most discussed issues in swimming hydrodynamics research.

Bixler and Riewald (2002) evaluated the steady flow around a swimmer's hand and forearm at various angles of attack and sweep back angles. The digital model was created based upon an adult male's right forearm and hand with the forearm fully pronated using similar procedures abovementioned. Force coefficients measured as a function of angle of attack showed that forearm drag was essentially constant and forearm lift was almost null. Additionally, hand drag presented the minimum value near angles of attack of 0° and 180° and the maximum value was obtained near 90°, when the model is nearly perpendicular to the flow. Hand lift was almost zero at 95° and presented the highest values near 60° and 150°. Regarding the water channel analysis, Schleihauf (1979) reported that lift coefficient values increased up to an attack angle around 40° and then decreased, although some differences with respect to the sweepback angle were observed. Drag coefficient values increased with increasing the attack angle and were less sensitive to sweepback angle changes.

Silva et al. (2008), using a real digital model of a swimmer hand and forearm, confirmed the supremacy of the drag component. They also revealed an important contribution of lift force to the overall propulsive force production by the hand/forearm in swimming phases, when the angle of attack is close to 45°. The drag coefficient presented higher values than the lift coefficient for all angles of attack. In fact, the drag coefficient increased with the angle of attack showing the maximum values with an angle of attack of 90° and the minimum values with an angle of attack of 0°. The lift coefficient of the model presented the maximum values with an angle of attack of 45°. Silva et al. (2008) obtained values of lift coefficient very similar for the angles of attack of 0° and 90°, although the minimum values were obtained with an angle of attack of 90°. Sato and Hino (2002) showed values of drag coefficient higher than lift coefficient at all angles of attack. From the results of the simulations the authors suggested that the resultant force was maximal with an angle of attack of 105° and the direction of the resultant force in that situation was -13°. Based on this analysis, the authors suggested stroke backward and with a little-finger-ward, out sweep motion, as the best stroke motion to produce the maximum thrust during underwater path.

Another interesting numerical report, when the sweep back angle is considered, was that more lift force is generated when the little finger leads the motion than when the thumb leads (Bixler & Riewald, 2002; Silva et al. 2008).

3.3 Contribution of arms and legs to propulsion

Another important issue is related to the contribution of arms and legs to propulsion. It is almost consensual that most propulsion is generated by the arms' actions. In front crawl swimming, it was found (Hollander et al., 1988; Deschodt, 1999) that about 85 to 90% of propulsion is produced by the arms' movements. Accordingly, the majority of the research under this scope is performed on arm's movements. Nevertheless, leg's propulsion should not be disregarded and future studies under this field should be addressed, helping swimmers to enhance performance. In this section it seems pertinent to present some interesting and pioneer studies of Lyttle and Keys (2006). These authors performed a three-dimensional numerical analysis, modelling the swimmer performing two kinds of underwater dolphin kick used after starts and turns, after the swimmer push-off from the wall: (i) high amplitude and low frequency dolphin kick and; (ii) low amplitude and high frequency dolphin kick. Main data demonstrated an advantage of using the large slow kick, over the small fast kick, concerning the velocity range that underwater dolphin kicks are used. In addition, changes were also made into the input kinematics (ankle plantar flexion angle) to demonstrate the practical applicability of the model. While the swimmer was gliding at 2.18 m/s, a 10° increase in ankle plantar flexion created greater propulsive force during the kick cycle. These results demonstrated that increasing angle flexibility would increase the stroke efficiency for the subject that was modelled, although some caution should be made when transferring this data into other swimmers with different anthropometrical profile.

Regarding arms' propulsion, Lecrivain et al. (2008) reported that the arm (and not only the hand and forearm) provided effective propulsion through most of the stroke, and this must be considered when studying the arm propulsion. In fact, Gardano and Dabnichki (2006) underlined the importance of the analysis of the entire arm rather than different parts of it. Thus, the authors concluded that drag profiles differed substantially with the elbow flexion angle, as the maximum value could vary by as much as 40%. In addition, Gardano and Dabnichki (2006) stated that maximum drag force was achieved by 160° of elbow angle. A prolonged plateau between 50° and 140° indicated greater momentum generated at 160° in comparison with the other configurations. This fact suggests a strong possibility for the existence of an optimal elbow angle for the generation of a maximum propulsive force. However, these findings are only possible to confirm if an entire model of the swimmer's arm, its movement relative to the body and the body's movement relative to the water are computed (Marinho et al., 2009c). This concern seems also an interesting topic to address in further studies.

3.4 Fingers relative position

Regarding arms' actions, a large inter-subject range of fingers relative position can be observed during training and competition, regarding thumb position and finger spreading. Due to the inherent inefficiency of human swimming, the question is: do any of these strategies enhance performance or is it just a more comfortable hand posture that swimmers assumed?

Regarding thumb position, although some differences in the results of different studies (Schleihauf, 1979; Takagi et al., 2001; Marinho et al., 2009a), main data seemed to indicate that when the thumb leads the motion (sweep back angle of 0°) a hand position with the thumb abducted would be preferable to an adducted thumb position. Additionally, Marinho et al. (2009a) found, for a sweep back angle of 0°, that the position with the thumb abducted presented higher values than the positions with the thumb partially abducted and adducted at angles of attack of 0° and 45°. At an angle of attack of 90°, the position with the thumb adducted presented the highest value of resultant force. Schleihauf (1979), using experimental procedures, found that the position with the thumb fully abducted showed a maximum lift coefficient at an angle of attack of 15°, whereas the models with partial thumb abduction showed a maximum value of lift coefficient at higher angles of attack (45°-60°). In these orientations, the position with the thumb partially abducted presented higher values than with the thumb fully abducted. Moreover, Takagi et al. (2001) also applying experimental measurements revealed that the thumb position influenced the lift force. For a sweep back angle of 0° (as used in the study of Marinho et al., 2009a) the model with abducted thumb presented higher values of lift force, whereas for a sweep back angle of 180° (the little finger as the leading edge), the adducted thumb model presented higher values of lift force. In addition, the drag coefficient presented similar values in the two thumb positions for a sweep back angle of 0° and higher values in the thumb adducted position for a sweep back angle of 180°.

Regarding different finger spreading, Marinho et al. (2010c), using a numerical analysis, studied the hand with: (i) fingers close together, (ii) fingers with little distance spread (a mean intra finger distance of 0.32 cm, tip to tip), and (iii) fingers with large distance spread (0.64 cm, tip to tip), following the same procedure when Schleihauf (1979) conducted his experimental research. Marinho et al. (2010c) found that for attack angles higher than 30°, the model with little distance between fingers presented higher values of drag coefficient when compared with the models with fingers closed and with large finger spread. For attack angles of 0°, 15° and 30°, the values of drag coefficient were very similar in the three models of the swimmer's hand. Moreover, the lift coefficient seemed to be independent of the finger spreading, presenting little differences between the three models. Nevertheless, Marinho et al. (2010c) were able to note slightly lower values of lift coefficient for the position with larger distance between fingers. In the same line of research, Minetti et al. (2009) showed, through numerical simulation of a three-dimensional model of the hand, that an optimal finger spacing (12°, roughly corresponding to the resting hand posture) increases the drag coefficient (+8.8%), which is 'functionally equivalent' to a greater hand palm area, thus a lower stroke frequency can produce the same thrust, with benefits to muscle, hydraulic and propulsive efficiencies. These results suggested that the hand seems to create more propulsive force when fingers are slightly spread. Flow visualization, through numerical simulations, provides an explanation for the increased force associated with the optimum finger spacing.

3.5 Steady vs. unsteady flow conditions

The majority of the abovementioned studies were conducted only under steady state flow conditions. However, one knows (Schleihauf, 1979) that swimmers do not move their arms/hands under constant velocity and direction motions. Therefore, some authors (Sanders, 1999; Bixler & Riewald, 2002; Sato & Hino, 2002; Rouboa et al., 2006) referred that it is important to consider unsteady effects when swimming propulsion is analysed. For instance, Bixler and Schloder (1996) analysed the flow around a disc with a similar area of a swimmer hand. Different simulations with different initial velocity and acceleration were conducted to model identical real swimming conditions, especially during insweep and upsweep phases of the front crawl stroke. According to the obtained results the authors reported that the hand acceleration could increase the propulsive force by around 24% compared with the steady flow condition. Sato and Hino (2002) using also numerical and experimental data showed that the hydrodynamic forces acting on the accelerating hand was much higher than with a steady flow situation and these forces amplifies as acceleration increases. Rouboa et al. (2006) analysed the effect of swimmer's hand/forearm acceleration on propulsive forces generation using numerical simulation techniques. The main data underlined that under the hand/forearm acceleration condition, the measured values for propulsive forces were approximately 22.5% higher than the forces produced under the steady flow condition. Thus, these data suggests that drag and lift forces produced by the swimmers' hand in a determined time are dependent not only on the surface area, the shape and the velocity of the segment but also on the acceleration of the propulsive segment.

3.6 Equipment

Research in this scope is very scarce and when applied the main focus is related to analyse how can different equipment improve swimming performance due to a decrease in hydrodynamic drag (Neiva et al., 2011). The study of the effects of using different swimsuits is a good example (Roberts et al., 2003; Pendergast et al., 2006), emphasising the importance of compression effects due to swimsuits on drag reduction (Neiva et al., 2011). However, to the best of our knowledge, the effects on improving propelling force had lower attention by swimming scientific community. Some equipment used by swimmers, especially during training, can be tested using numerical simulation techniques, allowing understand the specific effects on propelling force. For instance, the effects of wearing fins, and different types of fins (Tamura et al., 2002), paddles, and other devices using during training should be tested attempting to elucidate coaches to improve training efficiency.

4. Future research in swimming using numerical simulations

Throughout this chapter, several future ideas have been presented to improve the application of numerical simulations in swimming research. One of our major aims is to be able to evaluate biomechanical situations that can be used by coaches and swimmers to swim faster and, thus to enhance performance. Therefore, the effective evaluation of true swimming conditions should be a main focus. Under this scope, main concerns should be addressed to analyse unsteady flow conditions, studying arms and legs propulsion during actual swimming, adding body roll, movement of the body, rotations and accelerations of the propelling segments, on different swimming techniques.

As mentioned above, the analysis of the effects of different equipment and facilities on propelling forces seems to be an interesting and an important issue to be dealt in future studies.

5. Conclusion

During this chapter, the authors attempted to present some important studies that have been conducted in swimming research using numerical simulations. Although there are some limitations of these studies, it seems that this numerical tool should not be disregarded. Numerical simulations can be used to evaluate several hydrodynamic issues, hence helping swimmers moving faster. In the current work some issues regarding the effect of propelling force on swimming performance were discussed. The authors are aware of some limitations, although they believe that they were able to show the practical applications of numerical simulations to swimmers and their coaches.

Moreover, it was an attempt to address some concerns to be improved in future investigations.

6. Acknowledgement

This work was supported by the Portuguese Government by Grants of the Science and Technology Foundation (PTDC/DES/098532/2008).

7. References

Barbosa, T.M.; Bragada, J.A.; Reis, V.M.; Marinho, D.A.; Carvalho, C. & Silva, J.A. (2010). Energetics and biomechanics as determining factors of swimming performance: updating the state of the art. *Journal of Science and Medicine in Sports*, 13, 262-269

Bixler, B. & Schloder, M. (1996). Computational fluid dynamics: an analytical tool for the 21st century swimming scientist. *Journal of Swimming Research*, 11, 4-22.

Bixler, B.S. & Riewald, S. (2002). Analysis of swimmer's hand and arm in steady flow conditions using computational fluid dynamics. *Journal of Biomechanics*, 35, 713-717.

Bixler, B.; Pease, D. & Fairhurst, F. (2007). The accuracy of computational fluid dynamics analysis of the passive drag of a male swimmer. *Sports Biomechanics*, 6, 81-98.

Dabnichki, P. & Avital, E. (2006). Influence of the position of crew members on aerodynamics performance of two-man bobsleigh. *Journal of Biomechanics*, 39, 2733-2742.

Deschodt, V. (1999). Relative contribution of arms and legs in human to propulsion in 25 m sprint front crawl swimming. *European Journal of Applied Physiology*, 80, 192-199.

Fluent (2006). *Fluent 6.3 Documentation*. Fluent Inc., Hanover.

Gardano, P. & Dabnichki, P. (2006). On hydrodynamics of drag and lift of the human arm. *Journal of Biomechanics*, 39, 2767-2773.

Hinze, J.O. (1975). *Turbulence*. McGraw-Hill Publishing Co., New York.

Hollander, A.P.; de Groot, G.; van Ingen Schenau, G.; Kahman, R. & Toussaint, H. (1988). Contribution of the legs to propulsion in Front Crawl swimming. In: *Swimming Science V*, B. Ungerechts, K. Wilke & K. Reischle, (Eds.), 39-43, Human Kinetics Books, Illinois.

Kellar, W.P.; Pearse, S.R.G. & Savill, A.M. (1999). Formula 1 car wheel aerodynamics. *Sports Engineering*, 2, 203-212.

Lauder, M.; Dabnichki, P. & Bartlett, R. (2001). Improved accuracy and reliability of sweepback angle, pitch angle and hand velocity calculations in swimming. *Journal of Biomechanics*, 34, 31-39.

Lecrivain, G.; Slaouti, A.; Payton, C. & Kennedy, I. (2008). Using reverse engineering and computational fluid dynamics to investigate a lower arm amputee swimmer's performance. *Journal of Biomechanics*, 41, 2855-2859.

Liu, H. (2002). Computational biological fluid dynamics: digitizing and visualizing animal swimming and flying. *Integrative and Comparative Biology*, 42, 1050-1059.

Lyttle, A. & Keys, M. (2006). The application of computational fluid dynamics for technique prescription in underwater kicking. *Portuguese Journal of Sport Sciences*, 6, Suppl. 2, 233-235.

Marinho, D.A.; Rouboa, A.I.; Alves, F.B.; Vilas-Boas, J.P.; Machado, L.; Reis, V.M. & Silva, A.J. (2009a). Hydrodynamic analysis of different thumb positions in swimming. *Journal of Sports Science and Medicine*, 8, 1, 58-66.

Marinho, D.A.; Reis, V.M.; Alves, F.B.; Vilas-Boas, J.P.; Machado, L.; Silva, A.J. & Rouboa, A.I. (2009b). The hydrodynamic drag during gliding in swimming. *Journal of Applied Biomechanics*, 25, 3, 253-257.

Marinho, D.A.; Barbosa, T.M.; Kjendlie, P.L.; Vilas-Boas, J.P.; Alves, F.B.; Rouboa, A.I. & Silva, A.J. (2009c). Swimming simulation: a new tool for swimming research and practical applications. In: *Lecture Notes in Computational Science and Engineering – CFD for Sport Simulation*, M. Peters (Ed.), 33-62. Springer, Berlin.

Marinho, D.A.; Barbosa, T.M.; Kjendlie, P.L.; Mantripragada, N.; Vilas-Boas, J.P.; Machado, L.; Alves, F.B.; Rouboa, A.I. & Silva, A.J. (2010a). Modeling hydrodynamic drag in swimming using computacional fluid dynamics, In: *Computational Fluid Dynamics*, H.W. Oh (Ed.), 391-404. INTECH Education and Publishing, Vienna.

Marinho, D.A.; Reis, V.M.; Vilas-Boas, J.P.; Alves, F.B.; Machado, L.; Rouboa, A.I. & Silva, A.J. (2010b). Design of a three-dimensional hand/forearm model to apply Computational Fluid Dynamics. *Brazilian Archives of Biology and Technology*, 5(2), 437-442.

Marinho, D.A.; Barbosa, T.M.; Reis, V.M.; Kjendlie, P.L.; Alves, F.B.; Vilas-Boas, J.P.; Machado, L.; Silva, A.J. & Rouboa, A.I. (2010c). Swimming propulsion forces are enhanced by a small finger spread. *Journal of Applied Biomechanics*, 26, 87-92.

Minetti, A.E.; Machtsiras, G. & Masters, J.C. (2009). The optimum finger spacing in human swimming. *Journal of Biomechanics*, 42, 13, 2188-2190

Moreira, A.; Rouboa, A.; Silva, A.; Sousa, L.; Marinho, D.; Alves, F.; Reis, V.; Vilas-Boas, J.P.; Carneiro, A. & Machado, L. (2006). Computational analysis of the turbulent flow around a cylinder. *Portuguese Journal of Sport Sciences*, 6(Suppl. 1), 105.

Neiva, H.P.; Vilas-Boas, J.P.; Barbosa, T.M.; Silva, A.J. & Marinho, D.A. (2011). 13th FINA World Championships: Analysis of Swimsuits Used By Elite Male Swimmers. *Journal of Human Sport and Exercise*, 6, 1, 87-93.

Pallis, J.M.; Banks, D.W. & Okamoto, K.K. (2000). 3D computational fluid dynamics in competitive sail, yatch and windsurfer design, In: *The Engineering of Sport: Research, Development and Innovation*, F. Subic & M. Haake (Eds.), 75-79. Blackwell Science, Oxford.

Pendergast, D.R.; Capelli, C.; Craig Jr, A.B.; di Prampero, P.E.; Minetti, A.E.; Mollendorfl, J.; Termin, A. & Zamparo, P. (2006). Biophysics in swimming. *Portuguese Journal of Sport Sciences*, 6, Suppl. 2, 185-189.

Roberts, B.S.; Kamel, K.S.; Hedrick, C.E.; MLean, S.P. & Sharpe, R.L. (2003). Effect of Fastskin suit on submaximal freestyle swimming. *Medicine and Science in Sports and Exercise*, 35, 519-524.

Rouboa, A.; Silva, A.; Leal, L.; Rocha, J. & Alves, F. (2006). The effect of swimmer's hand/forearm acceleration on propulsive forces generation using computational fluid dynamics. *Journal of Biomechanics*, 39, 1239-1248.

Sanders, R.H. (1999). Hydrodynamic characteristics of a swimmer's hand. *Journal of Applied Biomechanics*, 15, 3-26.

Sanders, R.B.; Rushall, H., Toussaint, H., Steager, J. & Takagi, H. (2001). Bodysuit yourself but first think about it. *American Swimming Magazine*, 5, 23-32.

Sato, Y & Hino, T. (2002). Estimation of thrust of swimmer's hand using CFD. In: *Proceedings of 8th symposium on nonlinear and free-surface flows*, 71-75. Hiroshima.

Schleihauf, R.E. (1979). A hydrodynamic analysis of swimming propulsion. In: *Swimming III*, J. Terauds & E.W. Bedingfield (Eds.), 70-109. University Park Press, Baltimore.

Silva, A.J.; Rouboa, A.; Moreira, A.; Reis, V.; Alves, F.; Vilas-Boas, J.P. & Marinho, D. (2008). Analysis of drafting effects in swimming using computational fluid dynamics. *Journal of Sports Science and Medicine*, 7, 1, 60-66.

Takagi, H.; Shimizu, Y.; Kurashima, A. & Sanders, R. (2001). Effect of thumb abduction and adduction on hydrodynamic characteristics of a model of the human hand. In: *Proceedings of Swim Sessions of the XIX International Symposium on Biomechanics in Sport*, J. Blackwell & R. Sanders (Eds.), 122-126. University of San Francisco, San Francisco.

Tamura, H.; Nakazawa, Y.; Sugiyama, Y.; Nomura, T. & Torii, N. (2002). Motion analysis and shape evaluation of a swimming monofin. *The Engineering Sports*, 4, 716-724.

Toussaint, H. & Truijens, M. (2005). Biomechanical aspects of peak performance in human swimming. *Animal Biology*, 55, 1, 17-40.

4

Numerical Modeling and Simulations of Pulsatile Human Blood Flow in Different 3D-Geometries

Renat A. Sultanov and Dennis Guster
Department of Information Systems and BCRL,
St. Cloud State University, St. Cloud, MN
USA

1. Introduction

Cardiovascular diseases, such as ischemic heart disease, myocardial infarction, and stroke are leading causes of death in Western countries. All of these vascular diseases share a common element: atherosclerosis. They also share a common final event: the failure or destruction of the vascular wall structure, Dhein et al. (2005); Waite (2005).

Atherosclerosis reduces arterial lumen size through plaque formation and arterial wall thickening. It occurs at specific arterial sites. This phenomenon is related to hemodynamics and to wall shear stress (WSS) distribution, Fung (1993). From the physical point of view WSS is the tangential drag force produced by moving blood, i.e. it is a mathematical function of the velocity gradient of blood near the endothelial surface. A general description of WSS is presented in Landau & Lifshitz (1959). Arterial wall remodeling is regulated by WSS, Grotberg & Jensen (2004), for example, in response to high shear stress arteries enlarge. From the bio-mechanical point of view one can conclude, that the atherosclerotic plaques localize preferentially in the regions of low shear stresses, but not in regions of higher shear stresses. Furthermore, decreased shear stress induces intimal thickening in vessels which have adapted to high flow.

Final vascular events that induce fatal outcomes, such as acute coronary syndrome, are triggered by the sudden mechanical disruption of an arterial wall. Thus, we can conclude, that the final consequences of tragic fatal vascular diseases are strongly connected to mechanical events that occur within the vascular wall, and these in turn are likely to be heavily influenced by alterations in blood flow and the characteristics of the blood itself.

Currently researchers in the field of biomechanics and biomedicine conduct laboratory investigations of human blood flow in different shape and size tubes, which are designed to be approximate models of human vessels and arteries, see for example Huo & Kassab (2006); Taylor & Draney (2004). Some researchers also carry out intensive computer simulations of these bio-mechanical systems, see for example Chen & Lu (2004; 2006); Cho & Kensey (1991); Duraiswamy et al. (2007); Johnston et al. (2004); Morris et al. (2004; 2005); Mukundakrishnan et al. (2008); Peskin (1977); Sultanov et al., 2008 (a;b); Sultanov & Guster (2009).

Also, there have been laboratory experiments in which specific stents are incorporated in such artificial vessels (tubes). Stent implantation has been used to open diseased coronary blood vessels, allowing improved perfusion of the cardiac muscle. Used in combination with drug therapy, vascular repair and dilation techniques (angioplasty) the implantation of metallic stents has created a multibillion dollar industry. Stents are commonly used in many different blood vessels, but the primary site of deployment is in diseased coronary arteries.

Stents represent a very special case in the modeling research problems mentioned above, Frank et al. (2002). Taking into account that stents have a very small size and rather complicated structure and shape, this situation makes it difficult to obtain precise measurements. Therefore high quality and precise computer simulations of blood flow through vessels with implanted stents would be most useful, Frank et al. (2002). Work of this type is already underway and we would like to mention several pertinent studies, Benard et. al. (2006); Banerjee et. al. (2007); Faik et al. (2007); Seo et al. (2005).

Nevertheless, there are still many problems in obtaining precise realistic geometries for the required vessels. Human arteries, especially the aorta, have complicated spatial-geometric and characteristic configurations. For example, the aortic arch centerline does not lie on a plane and there are major branches at the top of the arch feeding the carotid arterial circulation. One of the main problems in the field of bio-medical blood flow simulation is to obtain precise geometrical-mathematical representations of different vessels. This information in turn needs to be included in the simulation programs.

However, it seems logical that a first step in these investigations would be to apply simpler 3D-geometry forms and models, but at the same time to take into account the precise physical effects of blood movement such as the non-Newtonian characteristics of human blood, realistic pulsatile flow, and possible turbulent effects. Because of the applied pulsatile flow in our simulations turbulence may be significant to the final results of this study.

Therefore, in the current work we carried out real-time full-dimensional computer simulations of a realistic pulsatile human blood flow in actual size vessels, vessels with a bifurcation, and in a model of the aortic arch. We take into account different physical effects, such as turbulence and the non-Newtonian nature of human blood. The next section presents the mathematical methodology and the physical model used in this work. The general purpose commercial computational fluid dynamics program FLOW3D is used for its basic functionality, but we supplemented its capability by adding our routines to obtain the results presented in this work.

Sec. 3 presents results for three vessels of different geometries. The CGS unit system is used in all simulations, as well as for presentation of the results. Conclusions and discussion comparing our results to well respected previous studies are included in Sec. 4.

2. Physical models and mathematical methods

As we mentioned above, we undertook pulsatile human blood flow simulation experiments using different size and shape human vessel arteries. For each spatial configuration one needs to provide a specific approach for the numerical solution to the complicated second order partial differential equations of the fluid dynamics. These equations are also named as the Navier-Stokes (NS) equations.

For simple cylindrical vessels we used the cylindrical coordinate system: $\vec{r} = (r, \theta, Z)$. However, for the aortic arch or bifurcated vessels, where there is no cylindrical symmetry, we applied the Cartezian coordinate system: $\vec{r} = (x, y, z)$. In the cases of the aortic arch and bifurcated vessels we used up to five blocks of matched Cartezian coordinate subsystems. Below we represent the NS equations in a general form, because, for each of the special cases, considered in this work and the chosen coordinate system, the partial differential equations of the fluid dynamics may look different. At the same time we understand that the general differential operator form of these equations is unique.

2.1 Equations

The general form of the dynamics equation for viscous fluid can be written in the following way, Landau & Lifshitz (1959):

$$\rho\left(\frac{\partial v_i}{\partial t} + v_k\frac{\partial v_i}{\partial x_k}\right) = -\frac{\partial p}{\partial x_i} + \frac{\partial}{\partial x_k}\left\{\eta\left(\frac{\partial v_i}{\partial x_k} + \frac{\partial v_k}{\partial x_i} - \frac{2}{3}\delta_{ik}\frac{\partial v_l}{\partial x_l}\right)\right\} + \frac{\partial}{\partial x_i}\left(\zeta\frac{\partial v_l}{\partial x_i}\right), \tag{1}$$

here v_i and x_k are velocity and coordinates of the fluid, ρ is the density of fluid, η and ζ are dynamical characteristics of the fluid, i.e. coefficients of viscosity. Because in a general case the pressure p, the temperature T and therefore the viscosity coefficients η and ζ are not constants in a flowing fluid, one cannot take them out of the partial differentials in the Eq. (1). However, in a very wide range of applications it is a good approximation to consider the variation of these coefficients to be negligible in the fluid, that is $\eta = const$ and $\zeta = const$. In these cases the Eq. (1) becomes the well known Navier-Stokes equation:

$$\rho\left[\frac{\partial \vec{v}}{\partial t} + (\vec{v}\nabla)\vec{v}\right] = -\mathrm{grad}\,p + \eta\triangle\vec{v} + (\zeta + \eta/3)\mathrm{grad}\,\mathrm{div}\vec{v}. \tag{2}$$

Further, if the fluid is considered as incompressible: div $\vec{v} = 0$, then the NS equation becomes simpler in form:

$$\left[\frac{\partial \vec{v}}{\partial t} + (\vec{v}\nabla)\vec{v}\right] = -\frac{1}{\rho}\mathrm{grad}\,p + \frac{\eta}{\rho}\triangle\vec{v}. \tag{3}$$

This equation is, probably, one of the most applicable mathematical results related to modeling and real time simulation of various physical systems, such as: air flow in aerodynamics, blood flow in medical applications and even cash flow in financial problems. The fundamental NS equation is nonlinear, diverse, rich, and, as we mentioned above, has strong practical applications in science and in various fields of modern technologies, for example, micro- and nano-fluidics. However, because of the considered atomistic level in the field of novel nano-fluidics problems a direct application of the NS equation to these systems might be problematic, especially at low temperatures of these systems. Further, from the mathematical point of view the NS equation is a very complicated nonlinear partial differential equation, which still remains as a "stumbling block" for mathematicians. The Clay Mathematics Institute (Boston, Massachusetts, USA) announced *Seven Millennium Problems* with a prize of US$ 1,000,000.00 for each (http://www.claymath.org/index.php). One of these problems is related to the existence and smoothness of the NS equation. It is hard to believe that despite many successful practical applications of the NS equation its fundamental mathematical property is still open to question.

Nevertheless, because this work deals with different 3D geometries it would be useful to represent the NS equation in different forms. For example, in the case of cylindrical symmetry one can apply cylindrical coordinates (r, θ, z) and the NS equation could be written:

$$\frac{\partial v_r}{\partial t} + v_r\frac{\partial v_r}{\partial r} + \frac{v_\theta}{r}\frac{\partial v_r}{\partial \theta} + v_z\frac{\partial v_r}{\partial z} - \frac{v_\theta^2}{r} = -\frac{1}{\rho}\frac{\partial p}{\partial r} + F_r + \nu\left(\nabla^2 v_r - \frac{v_r}{r^2} - \frac{2}{r^2}\frac{\partial v_\theta}{\partial \theta}\right), \tag{4}$$

$$\frac{\partial v_\theta}{\partial t} + v_r\frac{\partial v_\theta}{\partial r} + \frac{v_\theta}{r}\frac{\partial v_\theta}{\partial \theta} + v_z\frac{\partial v_\theta}{\partial z} + \frac{v_r v_\theta}{r} = -\frac{1}{\rho}\frac{1}{r}\frac{\partial p}{\partial \theta} + F_\theta + \nu\left(\nabla^2 v_\theta + \frac{2}{r^2}\frac{\partial v_r}{\partial \theta} - \frac{v_\theta}{r^2}\right), \tag{5}$$

$$\frac{\partial v_z}{\partial t} + v_r\frac{\partial v_z}{\partial r} + \frac{v_\theta}{r}\frac{\partial v_z}{\partial \theta} + v_z\frac{\partial v_z}{\partial z} = -\frac{1}{\rho}\frac{\partial p}{\partial z} + F_z + \nu\nabla^2 v_z, \tag{6}$$

together with the continuity equation:

$$\frac{1}{r}\frac{\partial}{\partial r}(rv_r) + \frac{1}{r}\frac{\partial v_\theta}{\partial \theta} + \frac{\partial v_z}{\partial z} = 0, \tag{7}$$

where $v = \eta/\rho$ is the kinematic viscosity Landau & Lifshitz (1959).

However, in the general case when there is no symmetry it is useful to apply the well known Cartesian coordinates x, y, and z: accordingly the equations of motion for the fluid velocity components (u, v, w) are:

$$\frac{\partial u}{\partial t} + u\frac{\partial u}{\partial x} + v\frac{\partial u}{\partial y} + w\frac{\partial u}{\partial z} = -\frac{1}{\rho}\frac{\partial p}{\partial x} + X + v\nabla^2 u, \tag{8}$$

$$\frac{\partial v}{\partial t} + u\frac{\partial v}{\partial x} + v\frac{\partial v}{\partial y} + w\frac{\partial v}{\partial z} = -\frac{1}{\rho}\frac{\partial p}{\partial y} + Y + v\nabla^2 v, \tag{9}$$

$$\frac{\partial w}{\partial t} + u\frac{\partial w}{\partial x} + v\frac{\partial w}{\partial y} + w\frac{\partial w}{\partial z} = -\frac{1}{\rho}\frac{\partial p}{\partial z} + Z + v\nabla^2 w. \tag{10}$$

In this case the continuity equation has the form:

$$\frac{\partial u}{\partial x} + \frac{\partial v}{\partial y} + \frac{\partial w}{\partial z} = 0. \tag{11}$$

In the specific case of the FLOW3D program the equations of motion for the fluid velocity components (u, v, w) with special additional terms included in the program are written:

$$\frac{\partial u}{\partial t} + \frac{1}{V_F}\left(uA_x\frac{\partial u}{\partial x} + vA_yR\frac{\partial u}{\partial y} + wA_z\frac{\partial u}{\partial z}\right) - \xi\frac{A_yv^2}{xV_f} = -\frac{1}{\rho}\frac{\partial p}{\partial x} + G_x + f_x - b_x - $$
$$\frac{R_{sor}}{\rho V_f}(u - u_w - \delta \cdot u_s) \tag{12}$$

$$\frac{\partial v}{\partial t} + \frac{1}{V_F}\left(uA_x\frac{\partial v}{\partial x} + vA_yR\frac{\partial v}{\partial y} + wA_z\frac{\partial v}{\partial z}\right) + \xi\frac{A_yuv}{xV_f} = -\frac{R}{\rho}\frac{\partial p}{\partial y} + G_y + f_y - b_y - $$
$$\frac{R_{sor}}{\rho V_f}(v - v_w - \delta \cdot v_s) \tag{13}$$

$$\frac{\partial w}{\partial t} + \frac{1}{V_F}\left(uA_x\frac{\partial w}{\partial x} + vA_yR\frac{\partial w}{\partial y} + wA_z\frac{\partial w}{\partial z}\right) = -\frac{1}{\rho}\frac{\partial p}{\partial z} + G_z + f_z - b_z - $$
$$\frac{R_{sor}}{\rho V_f}(w - w_w - \delta \cdot w_s). \tag{14}$$

Here, (u, v, w) are the velocity components in coordinate directions (x, y, z) respectively. For example, when Cartesian coordinates are used, $R = 1$ and $\xi = 0$, see FLOW3D manual FLOW3D (2007). A_x is the fractional area open to flow in the x direction, analogously for A_y and A_z. Next, V_F is the fractional volume open to flow, R and ξ are coefficients which depend on the coordinate system: (x, y, z) or (r, θ, z), ρ is the fluid density, R_{sor} is a mass source term. Finally, (G_x, G_y, G_z) are so called body accelerations FLOW3D (2007), (f_x, f_y, f_z) are viscous

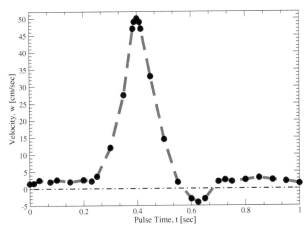

Fig. 1. The inflow time-dependent waveform used in these simulations. The function was taken from Fig. 3 of the work Papaharilaou et al. (2007).

accelerations, (b_x, b_y, b_z) are the flow losses in porous media or across porous baffle plates, and the final term accounts for the injection of mass at a source represented by a geometric component. Mathematical expressions for the viscous accelerations (f_x, f_y, f_z) are presented in the Appendix.

The term $U_w = (u_w, v_w, w_w)$ in equations (12-14) is the velocity of the source component, which will generally be non-zero for a mass source of a General Moving Object (GMO) FLOW3D (2007). The term $U_s = (u_s, v_s, w_s)$ is the velocity of the fluid at the surface of the source relative to the source itself. It is computed in each control volume as

$$\vec{U}_s = \frac{1}{\rho_s} \frac{d(Q\vec{n})}{dA} \tag{15}$$

where dQ is the mass flow rate, ρ_s fluid source density, dA the area of the source surface in the cell and $\vec{n}$ the outward normal to the surface. The source is of the stagnation pressure type when in equations (12-14) $\delta = 0.0$. Next, $\delta = 1.0$ corresponds to the source of the static pressure type.

As we mentioned earlier, in all simulations we considered the blood flow as a pulsatile flow. The final result for the inflow waveform has been taken from Fig. 3 of the work Papaharilaou et al. (2007). The pulse was applied for 5.5 cycle times in our work. The waveform is shown in Fig. 1. These velocity values are used as time-dependent inflow initial boundary conditions. These numbers are included directly in the FLOW3D program.

Next, the general mass continuity equation, which is solved within the FLOW3D program has the following *general* form:

$$V_f \frac{\partial \rho}{\partial t} + \frac{\partial}{\partial x}(\rho u A_x) + R \frac{\partial}{\partial y}(\rho v A_y) + \frac{\partial}{\partial z}(\rho w A_z) + \zeta \frac{\rho u A_x}{x} = R_{dif} + R_{sor}, \tag{16}$$

R_{dif} is a turbulent diffusion term, and R_{sor} is a mass source. The turbulent diffusion term is

$$R_{dif} = \frac{\partial}{\partial x}\left(v_p A_x \frac{\partial \rho}{\partial x}\right) + R \frac{\partial}{\partial y}\left(v_p A_y R \frac{\partial \rho}{\partial y}\right) + \frac{\partial}{\partial z}\left(v_p A_z \frac{\partial \rho}{\partial z}\right) + \zeta \frac{\rho v_p A_x}{x}, \tag{17}$$

where the coefficient $v_p = C_p \mu / \rho$, μ is dynamic viscosity and C_p is a constant. The R_{sor} term is a density source term that can be used to model mass injections through porous obstacle surfaces.

Compressible flow problems require the solution of the full density transport equation. In this work we treat blood as an incompressible fluid. For incompressible fluids $\rho = constant$, and the equation (15) becomes the following:

$$\frac{\partial}{\partial x}(uA_x) + \frac{\partial}{\partial y}(vA_y) + \frac{\partial}{\partial z}(wA_z) + \zeta \frac{uA_x}{x} = \frac{R_{sor}}{\rho}. \tag{18}$$

It is assumed, that at a stagnation pressure source fluid enters the domain at zero velocity. As a result, pressure should be considered at the source to move the fluid away from the source. For example, such sources are designed to model fluid emerging at the end of a rocket or the simple deflating process of a balloon. In general, stagnation pressure sources apply to cases when the momentum of the emerging fluid is created inside the source component, like in a rocket engine. At a static pressure source the fluid velocity is computed from the mass flow rate and the surface area of the source. In this case, no extra pressure is required to propel the fluid away from the source. An example of such a source is fluid emerging from a long straight pipe. Note that in this case the fluid momentum is created far from where the source is located.

Turbulence models can be taken into account in FLOW3D. It allows us to estimate the influence of turbulent fluctuations on mean flow quantities. This influence is usually expressed by additional diffusion terms in the equations for mean mass, momentum, and energy. The turbulence kinetic energy per unit mass, q, is the following:

$$\frac{\partial q}{\partial t} + \frac{1}{V_F}\left(uA_x\frac{\partial q}{\partial x} + vA_y R\frac{\partial q}{\partial y} + wA_z\frac{\partial q}{\partial z}\right) = P + G + Diff - D, \tag{19}$$

where P is shear production, G is buoyancy production, $Diff$ is diffusion, and D is a coefficient FLOW3D (2007).

When the turbulence option is used, the viscosity is a sum of the molecular and turbulent values. For non-Newtonian fluids the viscosity can be a function of the strain rate and/or temperature. A general expression based on the Carreau model is used in FLOW-3D for the strain rate dependent viscosity:

$$\mu = \mu_\infty + \frac{\mu_0 E_T - \mu_\infty}{\lambda_{00} + [\lambda_0 + (\lambda_1 E_T)^2 e_{ij}e_{ij}]^{(1-n)/2}} + \frac{\lambda_2}{\sqrt{(e_{ij}e_{ij})}}, \tag{20}$$

where $e_{ij} = 1/2(\partial u_i/\partial x_j + \partial u_j/\partial x_i)$ is the fluid strain rate in Cartesian tensor notations, $\mu_\infty, \mu_0, \lambda_0, \lambda_1, \lambda_2$ and n are constants. Also, $E_T = exp[a(T^*/(T - b) - C)]$, where T^*, a, b, and C are also parameters of the temperature dependence, and T is fluid temperature. This basic formula is used in our simulations for blood flow in vessels and in the aortic arch. For a variable dynamic viscosity μ, the viscous accelerations have a special form. That form is shown in the Appendix.

The equations of fluid dynamics should be solved together with specific boundary conditions. The numerical model starts with a computational mesh, or grid. It consists of a number of interconnected elements, or 3D-cells. These 3D-cells subdivide the physical space into small volumes with several nodes associated with each such volume. The nodes are used to store values of the unknown parameters, such as pressure, strain rate, temperature, velocity components and etcetera. This procedure provides values for defining the flow parameters at discrete locations and allows specific boundary conditions to be set up. As the culminating step, one can start developing effective numerical approximations for the solution of the fluid dynamics equations, i.e. NS equation.

New pressure-velocity solutions have been implemented in FLOW-3D. We used the GMRES method. GMRES stands for the generalized minimum residual method. In addition to the GMRES solution, a new optional algorithm, the generalized conjugate gradient (GCG) algorithm, has also been implemented for solving viscous terms in the new GMRES routine. This new solver is a highly accurate and efficient method for a wide range of problems. It possesses good convergence, symmetry and speed properties; however, it does use more memory than the SOR or SADI methods.

3. Numerical results

Results of our simulations are presented below. One of the most important preliminary testing tasks is to check for numerical convergence. This test has been successfully accomplished in this work. A portion of the test calculation results are shown below in this paper. Next, in this work particular attention has been given to the calculations of the wall shear stress distribution (WSS). As we mentioned above WSS is the tangential drag force produced by moving blood, i.e. it is a mathematical function of the velocity gradient of blood near the endothelial surface:

$$\tau_w = \mu \left[\frac{\partial U(t, y, R_v)}{\partial y} \right]_{y \approx 0}. \tag{21}$$

Here μ is the dynamic viscosity, t is current time, $U(t, y, R_v)$ is the flow velocity parallel to the wall, y is the distance to the wall of the vessel, and R_v is its radius. It was shown, that the magnitude of WSS is directly proportional to blood flow and blood viscosity and inversely proportional to the cube of the radius of the vessel, in other words a small change of the radius of a vessel will have a large effect on WSS.

First, we present results for a simple geometry vessel in the shape of a tube. However, the human blood is treated as real and a non-Newtonian liquid. The necessary data for viscosity of the blood was found in previous laboratory and clinical measurements.

As we mentioned above, we take into account the real pulsatile flow, which is shown in Fig. 1. The data for Fig. 1 have also been obtained in clinical measurements, Papaharilaou et al. (2007). After such preliminary simulations we switched to a more complicated spatial configurations. This work begins with coronary bifurcation and the aortic arch. It is axiomatic that real people may have different size aortic arches with slightly different shapes. However, we carried out simulations for an average size and shape aortic arch.

The main goal of this work is to treat the above mentioned systems realistically, reveal the physics of the blood flow dynamics, and to obtain reliable results for pressure, dynamic viscosity, velocity profiles and strain rate distributions. Also, we tested the widely cited Newtonian and non-Newtonian models of the human blood.

3.1 Straight vessel: cylinder

First, we selected a simple vessel geometry, that is we considered the shape of a straight vessel to be a tube. In our simulations involving a straight cylinder type vessel we applied a cylindrical coordinate system: (r, θ, Z) with the axis OZ directed over the tube axis. Different quantities of cells have been used to discretize the empty space inside the tube. In the open space (inner part of the tube) the fluid dynamics equations have been solved using appropriate

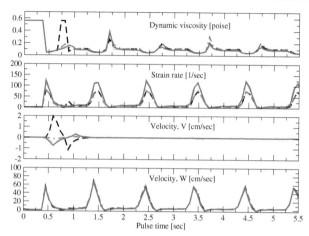

Fig. 2. Test of numerical convergence. Time-dependent dynamic viscosity, strain rate and velocity components V and W. Results for a vessel of simple geometry - cylinder type, for a specific spatial point inside the cylinder - the middle point. No turbulence effects are involved in these simulations with the realistic non-Newtonian viscosity of human blood. Black dashed line: calculations with 0.08 size for all cells FLOW3D (2007), red dot-dashed line with 0.07, green double dot - dashed line with 0.065, and blue bold line calculations with 0.062 size for all cells.

mathematical boundary conditions. The size of the tube is: $L = 8$ cm (in length) and $R = 0.34$ cm (length of inner radius). The thickness of the vessel wall is $s = 0.03$ cm. We have applied 5.5 cycles of blood pulse.

Let us now evaluate the expression (20). In these calculations we followed the work Cho & Kensey (1991), where the Carreau model of the human blood has also been used. To be consistent with Cho & Kensey (1991) we choose the following coefficients: $\lambda_2 = \lambda_{00} = 0, a = 0$ and $E_T = 1$, that is we don't take into account the temperature dependence of the viscosity. This investigation is to be addressed in our subsequent work. Next: $\lambda_0 = 1, \lambda_1 = 3.313$ sec, $\mu_\infty = 0.0345$ P, $\mu_0 = 0.56$ P, and $n = 0.3568$. The convergence was achieved when we used 52,800 cells, that is we used 100 points over OZ, 22 points over the radius of the inside space $R = 0.34$ cm, and 24 points over azimuthal angle Φ from 0 to 2π.

Time-dependent results for dynamic viscosity, strain rate and velocity components V and W are presented in Fig. 2. The turbulent effects are not taken into account. We decided to present only one precise geometrical point for comparison purposes: the middle point: $r = \theta = 0$, and $Z = 4.0$ cm. The data for Fig. 2 were obtained with the non-Newtonian model of human blood. We refer the reader to the comments provided for the figure. We were able to closely replicate the values for all previous cell sizes FLOW3D (2007) and obtain almost identical values, for example for pressure, wall shear stress and other parameters, for 0.065 mm and 0.062 mm cell sizes FLOW3D (2007). This means, that the convergence has been achieved.

Next, it would be very interesting to compare the results calculated with and without the turbulent effect. To support this endeavor we used the realistic non-Newtonian model of blood viscosity, the pulsatile flow, and the size of computation cells at which convergence has been achieved, that is the 0.062 mm size for all computational cells FLOW3D (2007). The results are presented in Fig. 3. As we see from Fig. 3 the effect of the turbulence is

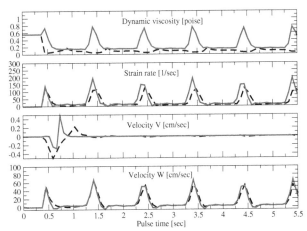

Fig. 3. Time-dependent results for a specific geometrical point inside the cylinder: the middle point. Black dashed line: simulations without taking into account the turbulence; red bold line results with the turbulence. The non-Newtonian viscosity is taken into account.

significant, particularly in regard to dynamic viscosity and strain rate. This result means that in the case of pulsatile flows and non-Newtonian viscosity the turbulent term should be taken into account. In Fig. 4 we separately show the results for the pulsatile pressure distribution and the turbulent energy, again using the middle point of the cylinder.

Finally, it would also be very interesting to make a comparison between the results calculated using both a Newtonian and non-Newtonian viscosity. However, as in previous simulations, we will apply the pulsatile flow with the turbulence included, since it has proved to be important. The results are shown in Fig. 5. As one can see, we obtain significant differences between these two calculations. We specifically observed that for the pressure distribution, dynamic viscosity and turbulent energy, we obtained significant variation.

Thus, we arrive at the important conclusion: within a time-dependent (pulsatile) flow of human blood it is necessary to take into account turbulence and non-Newtonian viscosity. The bold lines are results with the non-Newtonian viscosity (20) and the dashed lines are results with the Newtonian model when the viscosity μ has a constant value and is equal to 0.0345 P. As one can see the results are different for strain rate distributions and very different for pressure distributions. These results clearly indicate that in most cases when computer simulations are used in regard to human blood flow only the non-Newtonian model should be used.

3.2 Hemodynamics in the coronary bifurcation

Below we show the result of a subsequent simulation involving a $90°$ bifurcated coronary artery in Figs. 6 and 7. The geometrical model of the bifurcation consisted of a $90°$ intersection of two cylinders. This model represents the bifurcation between the left anterior descending coronary artery and the circumflex coronary artery. In our opinion, in the case of pulsatile flow it is more interesting to present results in a time-dependent way. This method can provide a wider picture of highly non-stationary flow systems. In this paper, because of space limitations, we just included time-dependent results for pressure, dynamic viscosity,

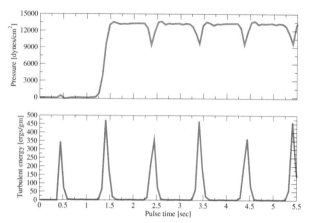

Fig. 4. Time-dependent results for pressure and the turbulent energy in the middle point of the cylinder. The non-Newtonian viscosity.

turbulent energy, and strain rate. However, we understand, that results which depend on spatial coordinates (r, θ, Z) for a few fixed moments of time are also highly useful.

In the case of the bifurcation shown in Fig. 6, we report the results for only two spatial points, which are the two outflow sides: the far right side and the farthest upper side of the bifurcation. The length of the lower horizontal vessel is 4 cm and its diameter is 0.54 cm. The length of the upper vertical vessel is 1.2 cm and its diameter is 0.4 cm. These sizes are consistent with average size human vessels.

Further, Fig. 6 represents our time-dependent results for the two outflow sides mentioned above. These results are for pressure, dynamic viscosity, turbulent energy and strain rate. The bold black lines are the results for the right outflow side, and the red dashed lines are the results for the farthest upper side (see comments to Fig. 6). In conclusion, the main goal of these calculations is to adopt them to investigate a case in which a stent is implanted in the bifurcation area Frank et al. (2002).

In Fig. 7 blood flows in from the left to the right with the imposed initial velocity profile taken from Fig. 1. The pressure, strain rate and turbulent energy distributions are shown for only one specific time moment $t=4.329$ s. The velocity vectors are also shown on these plots.

3.3 Blood flow in aortic arch

The geometry of the blood simulations inside the human aortic arch is shown in Figs. 8 and 9. On the top of the aortic arch three arteries are included. These arteries deliver the blood to the carotid artery and then to the brain. This configuration only models and approximately represents the real aortic arch. One of the goals of our simulations is to reveal the physics of the blood flow dynamics in this important portion of the human cardiovascular system.

The aortic arch is represented as a curved tube. The outer radius of the tube is 2.6 cm. A straight vessel (tube) is also merged to the arch. The length of the straight tube is about 4 cm. Again, the thickness of the wall is 0.03 cm, and the inner radius of the tube is $r = 0.34$ cm. The thickness is not important in these simulations, but it will be useful when, in future works, we will need to introduce elasticity of the walls of the tubes. The FLOW3D program allows to carry out fluid dynamic simulations with elastic (not only hard body) walls.

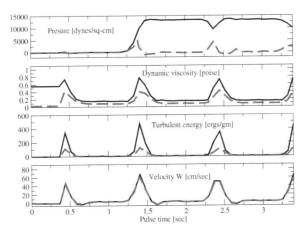

Fig. 5. Results for pressure, dynamic viscosity, turbulent energy and velocity W.
Time-dependent results for the middle point of the cylinder. Bold black line calculations with
non-Newtonian viscosity of the human blood; red dashed line with its Newtonian
approximation.

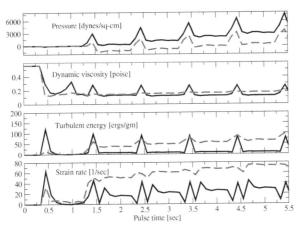

Fig. 6. Time-dependent results for a vessel with bifurcation. Pulsatile blood flow,
non-Newtonian viscosity, and the turbulence effect is included. Bold black line: results for
the far right outflow side $z = 0.0$; red dashed line results for the farthest up outflow side
$y = 0.0$. Convergence test results for velocity components: U, V, W.

Once again we are using the Cartesian coordinate system. We also carried out a convergence
test. To better represent the shape of the arch we applied five Cartesian sub-coordinate
systems in our FLOW3D simulations. After the discretization the total number of all cubic
cells reached about 900,000. It is important that once again we obtained full numerical
convergence. Again, the geometry is shown in Figs. 8 and 9. In this work we computed
pressure, velocity and strain rate distributions in the arch, while the human blood is treated
as a non-Newtonian liquid and while the realistic pulsatile blood flow is used.

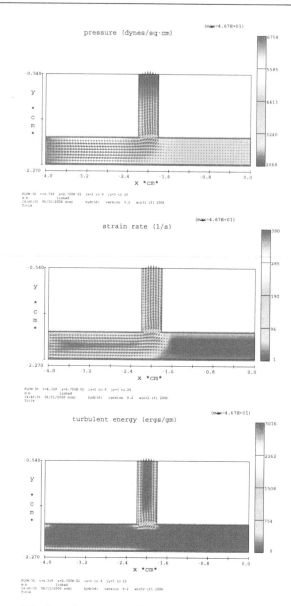

Fig. 7. The figures are 2D-plots showing the blood flow in the bifurcated vessels for only one precise moment of the discretized time t_i = 4.329 sec, the corresponding index is i = 40. Upper plot represents the result for the pressure distribution in the bifurcation, and the pressure ranges from 2068 dynes/sq-cm to 6758 dynes/sq-cm. The middle plot represents the results for the strain rate distribution and the lower plot shows results for the turbulent energy in the bifurcation. The range of the values is also shown.

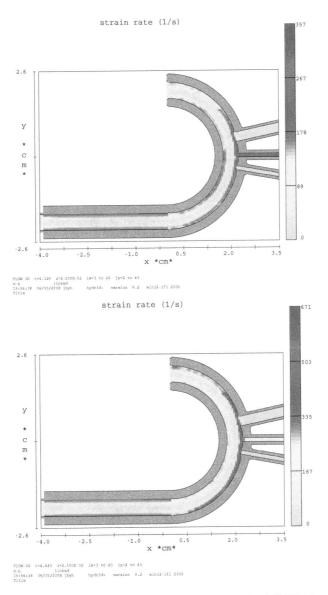

Fig. 8. Blood flow in the aortic arch. These two plots represent the full 2D-picture of the geometry used in these simulations. Shaded results for the strain rate are also shown, the bars on the right show the values. Results are for two specific moments of the time $t_{40} = 4.329$ sec and $t_{41} = 4.440$ sec. The values of the strain rate distribution range from 0.0 1/sec to 357.0 1/sec (upper plot) and from 0.0 to 671 1/sec (lower plot). The maximum values of the strain rate are localized in the region inside the arch. Blood flows from right to left in both pictures.

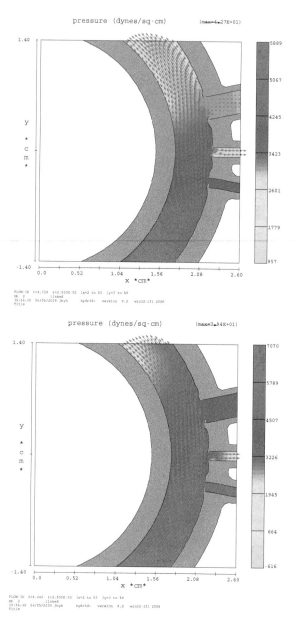

Fig. 9. Blood flow in the aortic arch. These two plots represent in more detail the region of the arch together with shaded results for the pressure distribution. The bars on the right show the values. These results are for two specific moments of the time $t_{40} = 4.329$ sec and $t_{41} = 4.440$ sec, where the pressure ranges from 957 dynes/sp-cm to 5889 dynes/sq-cm (upper plot), and from -616 dynes/sq-cm to 7070 dynes/sq-cm (lower plot).

In Fig. 8 we present the results of strain rate distributions inside the arch for two specific time moments. At the most left point, which is the inlet, we specify the pulsatile velocity source as the initial condition, that is the data from Fig. 1 are used. From the general theory of fluid mechanics Landau & Lifshitz (1959) it is possible to determine together with the blood density, viscosity, and spatial geometries, the dynamics of the blood according to the Navier-Stokes equation and its boundary conditions. Small vectors indicate the blood velocity. As can be seen from Fig. 8 blood flows from left to right in direction. However, because of pulsatility blood flows in the opposite direction too.

The values of the strain rate are also shown. These values are strongly oscillating. From the plots one can conclude that in the region of the arch the strain rate values become much larger than in the region of the straight vessel. This result represents clear evidence that in this part of the human vascular system atherosclerotic plaques should localize less than in the straight vessels. However, the higher wall shear stress values in the aortic arch could be the reason for sudden mechanical disruption of the arterial wall in this part of the human vascular system. These results are consistent with laboratory and clinical observations. In Fig. 9 we depict the pressure distribution in the arch.

4. Conclusion

In this work we applied computational fluid dynamics techniques to support pulsatile human blood flow simulations through different shape/size vessels and the aortic arch. The realistic blood pulse has been adopted and applied from the work Papaharilaou et al. (2007). The geometrical size of the vessels and the aortic arch have been selected to match the average real values. Human blood was treated in two different ways: (a) as a Newtonian liquid when the viscosity of the blood has a constant value, and (b) as a non-Newtonian liquid with the viscosity value represented by the equation (20). The numerical coefficients in (20) have been taken from work, Cho & Kensey (1991).

It is always difficult to obtain a steady-state cycle profile and stable computational results at the very beginning of time-dependent simulations. However, after a short stabilization period a steady-state cycle profile can be obtained. In our simulations we used up to 5.5 pulse cycles to reach complete steady state profiles. We obtained valid results for pressure, wall shear stress distribution and other physical parameters, such as the three velocity components of blood flow. All of these were shown in Figs. 2-6.

Our simulations showed that the FLOW3D program is capable of providing stable numerical results for all geometries included in this work. The time-dependent mathematical convergence test has been successfully carried out. Particular attention has been paid to this aspect of the calculations. It is a well known fact that fluid dynamics equations can have unstable solutions, Landau & Lifshitz (1959). Therefore, numerical convergence has been tested and confirmed in this work.

The result of computer simulations of blood flow in vessels for three different geometries have been presented. For pressure, strain rate and velocity component distributions we found significant disagreements between our results obtained with the realistic non-Newtonian treatment of human blood and the widely used method in literature: a simple Newtonian approximation.

Our results are in good agreement with the conclusions of the works, Chen & Lu (2004; 2006), where the authors also obtained significant differences between their results calculated with and without the non-Newtonian effect of blood viscosity. However, the recent work, Boyd & Buick (2007) should be mentioned, in which the authors performed 2-dimensional simulations of human blood flow through the carotid artery with and without the non-Newtonian effect of the viscosity. They did not find any substantial differences in their results. Finally, we would like to mention the paper Agarwal et. al. (2008), where the authors also performed simulations for the carotid artery, but only the non-Newtonian viscosity was used.

Next, the influence of a possible turbulent effect has also been investigated in this work. It was found that the effect is important. We believe, that the physical reason of this phenomena lies in the strong pulsatility of the flow and in the non-Newtonian viscosity of the blood. The contribution of the turbulence is most significant in the area of bifurcated vessels.

Finally, a significant increase of the strain rate and, the wall shear stress distribution, is found in the region of the aortic arch. This computational result provides additional evidence to support recent clinical and laboratory observations that this part of the human cardiovascular system is under higher risk of disruption Carter et al. (2001); Pochettino & Bavaria (2006). In future work it would be interesting to include the elasticity of the walls of the aortic arch Fang et al. (1998) and other vessels.

In conclusion, we would like to specifically point out, that the developments in this work can be directly applied to even more interesting and very important situations such as when a stent is implanted inside a vessel Frank et al. (2002). In this case, for example, it would be very useful to determine blood flow disturbance, the pressure distribution, strain rate and values of other physical parameters. The results of this work should allow us to determine the optimal size and shape of effective stents. As we mentioned in the Introduction some research groups are carrying out laboratory and computer simulations of blood flow through vessels with implanted stents Frank et al. (2002). It is very difficult to underestimate the value of these works.

5. Acknowledgments

This work was partially supported by Office of Sponsored Programs (OSP), by Internal Grant Program of St. Cloud State University, St. Cloud, MN-56301-4498, USA, and by a private Minnesota based company: RIE Coatings, Eden Valley, MN-55329-1646, USA (www.riecoatings.com).

6. Appendix

For a variable dynamic viscosity μ, the viscous accelerations are

$$\rho V_F f_x = w_x^s - [\frac{\partial}{\partial x}(A_x \tau_{xx}) + R\frac{\partial}{\partial y}(A_y \tau_{xy}) + \frac{\partial}{\partial z}(A_z \tau_{xz}) + \frac{\xi}{x}(A_x \tau_{xx} - A_y \tau_{yy})] \qquad (22)$$

$$\rho V_F f_y = w_y^s - [\frac{\partial}{\partial x}(A_x \tau_{xy}) + R\frac{\partial}{\partial y}(A_y \tau_{yy}) + \frac{\partial}{\partial z}(A_z \tau_{yz}) + \frac{\xi}{x}(A_x + A_y \tau_{xy})] \qquad (23)$$

$$\rho V_F f_z = w_z^s - [\frac{\partial}{\partial x}(A_x \tau_{xz}) + R\frac{\partial}{\partial y}(A_y \tau_{yz}) + \frac{\partial}{\partial z}(A_z \tau_{zz}) + \frac{\xi}{x}(A_x \tau_{xz})], \qquad (24)$$

where

$$\tau_{xx} = -2\mu \left(\frac{\partial u}{\partial x} - \frac{1}{3} \left(\frac{\partial u}{\partial x} + R\frac{\partial v}{\partial y} + \frac{\partial w}{\partial z} + \frac{\xi u}{x} \right) \right) \qquad (25)$$

$$\tau_{yy} = -2\mu \left[R\frac{\partial v}{\partial x} + \xi\frac{u}{x} - \frac{1}{3} \left(\frac{\partial u}{\partial x} + R\frac{\partial v}{\partial y} + \frac{\partial w}{\partial z} + \frac{\xi u}{x} \right) \right] \qquad (26)$$

$$\tau_{zz} = -2\mu \left(\frac{\partial w}{\partial z} - \frac{1}{3} \left(\frac{\partial u}{\partial x} + R\frac{\partial v}{\partial y} + \frac{\partial w}{\partial z} + \frac{\xi u}{x} \right) \right) \qquad (27)$$

$$\tau_{xy} = -\mu \left(\frac{\partial v}{\partial x} + R \frac{\partial u}{\partial y} - \frac{\xi v}{x} \right) \tag{28}$$

$$\tau_{xz} = -\mu \left(\frac{\partial u}{\partial z} + \frac{\partial w}{\partial x} \right) \tag{29}$$

$$\tau_{yz} = -\mu \left(\frac{\partial v}{\partial z} + R \frac{\partial w}{\partial y} \right). \tag{30}$$

In the above equations (22)-(24) the terms w_x^s, w_y^s and w_z^s are wall shear stresses. If these terms are equal to zero, there is no wall shear stress. This is because the remaining terms contain the fractional flow areas (A_x, A_y, A_z) which vanish at the walls FLOW3D (2007).

The wall stresses are modeled by assuming a zero tangential velocity on the portion of any area closed to flow. Mesh and moving obstacle boundaries are an exception because they can be assigned non-zero tangential velocities. In this case the allowed boundary motion corresponds to a rigid body translation of the boundary parallel to its surface. For turbulent flows, a law-of-the-wall velocity profile is assumed near the wall, which modifies the wall shear stress magnitude.

7. References

Agarwal, R., Katiyar, V.K. & Pradhan, P. (2008). A mathematical modeling of pulsatile flow in carotid artery bifurcation, Intern. J. of Engineering Science 46 (11), pp. 1147-1156.

Benard, N., Perrault, R. & Coisne, D. (2006). Computational Approach to Estimating the Effects of Blood Properties on Changes in Intra-Stent Flow, Annals of Biomed. Engineering 34 (8), pp. 1259-1271.

Banerjee, R.K., Devarakonda, S.B., Rajamohan, D. & Back, L.H. (2007). Developed Pulsatile Flow in a Deployed Coronary Stent, Biorheology 44 (2), pp. 91-102.

Boyd, J. & Buick, J.M. (2007). Comparison of Newtonian and non-Newtonian Flows in a Two-Dimensional Carotid Artery Model Using the Lattice Boltzmann Method, Physics in Medicine and Biology 52(20), pp. 6215-6228.

Carter, Y.M, Karmy-Jones, R.C., Oxorn, D.C. & Aldea, G.S. (2001).Traumatic Disruption of the Aortic Arch, European J. Cardiothoracic Surgery, 20, pp. 1231.

Chen, J. & Lu, X.-Y. (2005). Numerical investigation of the non-Newtonian blood flow in a bifurcation model with a non-planar branch Journal of Biomechanics, 37 (12), pp. 1899-1911.

Chen, J. & Lu, X.-Y. (2006). Numerical investigation of the non-Newtonian pulsatile blood flow in a bifurcation model with a non-planar branch, Journal of Biomechanics 39 (5), pp. 818-832.

Cho, Y.I. & Kensey, K.R. (1991). Effects of the non-Newtonian viscosity of blood on flows in a diseased arterial vessel. Part 1: Steady Flows, Biorheology 28(3-4), pp. 241-262.

Dhein, S., Delmar, M., & Mohr, F.W. (2005). Practical Methods in Cardiovascular Research, Springer-Verlag New-York, LLC.

Duraiswamy, N., Schoephoerster, R.T., Moreno, M.R. & Moore Jr., J.E. (2007). Stented Artery Flow Patterns and Their Effects on the Artery Wall, Ann. Rev. of Fluid Mechanics 39, pp. 357-382.

Faik, I., Mongrain, R., Leask, R.L., Rodes-Cabau, J., Larose, E. & Bertrand, O. (2007). Time-Dependent 3D Simulations of the Hemodynamics in a Stented Coronary Artery, Biomedical Materials 2 (1), art. no. S05, S28-S37.

Fang, H., Lin,Z. & Wang,Z. (1998). Lattice Boltzmann Simulation of Viscous Fluid Systems with Elastic Boundaries, Phys. Rev. E 57(1), R25-R28.

FLOW-3D Users Manual. (2007). Version 9.2, Flow Science, Santa Fe, New Mexico, USA.

Frank, A.O., Walsh, P.W., & Moore Jr., J.E. (2002). Computational Fluid Dynamics and Stent Design, Artificial Organs 26(7), pp. 614-621.

Fung, Y.C. (1993). Biomechanics, Springer.

Grotberg, J.B. & Jensen, O.E. (2004). Biofluid Mechanics in Flexible Tubes, Ann. Rev. Fluid Mech. 36, pp. 121-147.

Huo, Y. & Kassab, G.S. (2006). Pulsatile Blood Flow in the Entire Coronary Arterial Tree: Theory and Experiment, Am. J. Physiol. Heart Circ. Physiol 291(3) pp. H1074-H1087.

Johnston, B.M., Johnston, P.R., Corney, S. & Kilpatrick, D. (2004). Non-Newtonian Blood Flow in Human Right Coronary Arteries: Steady State Simulations, J. of Biomech. 37(5), pp. 709-720.

Landau, L.D. & Lifshitz, E.M. (1959). Fluid Mechanics, Volume 6, Pergamon Press Ltd.

Morris, L., Delassus, P., Walsh, M. & McGloughlin, T. (2004). A mathematical model to predict the in vivo pulsatile drag forces acting on bifurcated stent grafts used in endovascular treatment of abdominal aortic aneurysms (AAA): J. of Biomechanics 37(7), pp. 1087-1095.

Morris, L., Delassus, P., Callanan, A., Walsh, M., Wallis, F., Grace, P. & McGoughlin, T. (2005). 3-D Numerical Simultation of Blood Flow Throught Models of the Human Aorta, J. of Biomech. Engineering 127(5), pp. 767-775.

Mukundakrishnan, K., Ayyaswamy, P.S. & Eckmann, D.M. (2008). Finite-sized gas bubble motion in a blood vessel: Non-Newtonian effects, Phys. Rev. E 78(3), art. no. 036303.

Papaharilaou, Y., Ekaterinaris, J.A., Manousaki, E. & Katsamouris, A.N. (2007). A Decoupled Fluid Structure Approach for Estimating Wall Stress in Abdominal Aortic Aneurysms, J. Biomechanics 40(2), pp. 367-377.

Peskin, C.S. (1977). Numerical Analysis of Blood Flow in the Heart, Journal of Computational Physics, 25 (3) pp. 220-252.

Pochettino, A. & Bavaria, J.E. (2006). Aortic Dissection, in Book: Mastery of Cardiothoracic Surgery, Eds. L.R. Kaiser, I.L. Kron, T.L. Spray, Publisher: Lippincott Williams and Wilkins, pp. 534-544.

Seo, T., Schachter, L.G. & Barakat, A.I. (2005). Computational Study of Fluid Mechanical Disturbance Induced by Endovascular Stents, Annals of Biomed. Engineering 33 (4), pp. 444-456.

Sultanov, R.A., Guster, D., Engelbrekt, B. & Blankenbecler, R. (2008). A Full Dimensional Numerical Study of Pulsatile Human Blood Flow in Aortic Arch, Proceedings of the 2008 International Conference on Bioinformatics and Computational Biology, Vol. 2, pp. 437-443, Eds. H.M. Arabnia, M.Q. Yang, J.Y. Yang, CSREA Press-WORLDCOMP;

Sultanov, R.A., Guster, D., Engelbrekt, B., & Blankenbecler, R. (2008). 3D Computer simulations of pulsatile human blood flows in vessels and in aortic arch: investigation of non-Newtonian characteristics of human blood, Proceedings of the 2008 11-th IEEE International Conference on Computational Science and Engineering, art. no. 4578268, pp. 479-485, IEEE Comp. Soc.

Sultanov, R. A. & Guster, D. (2009). Full dimensional computer simulations to study pulsatile blood flow in vessels, aortic arch and bifurcated veins: Investigation of blood viscosity and turbulent effects, Proceedings of the 31st Annual International Conference of the IEEE Engineering in Medicine and Biology Society: Engineering the Future of Biomedicine, EMBC 2009 , art. no. 5334202, pp. 4704-4710.

Taylor, C.A. & Draney, M.T. (2004). Experimental and Computational Methods in Cardiovascular Fluid Mechanics, Annual Review of Fluid Mechanics 36, pp. 197-231.

Waite, L. (2005). Biofluid Mechanics in Cardiovascular Systems, Mc-Graw-Hill Professional Publishing.

5

Assessment of Carotid Flow Using Magnetic Resonance Imaging and Computational Fluid Dynamics

Vinicius C. Rispoli[1], Joao L. A. Carvalho[1], Jon F. Nielsen[2]
and Krishna S. Nayak[3]
[1]*Universidade de Brasília*
[2]*University of Michigan*
[3]*University of Southern California*
[1]*Brazil*
[2,3]*USA*

1. Introduction

Knowledge of blood flow patterns in the human body is a critical component in cardiovascular disease research and diagnosis. Carotid atherosclerosis, for example, refers to the narrowing of the carotid arteries. One symptom of atherosclerosis is abnormal flow. The carotid arteries supply blood to the brain, so early detection of carotid stenosis may prevent thrombotic stroke. The current clinical gold standard for cardiovascular flow measurement is Doppler ultrasound. However, evaluation by ultrasound is inadequate when there is fat, air, bone, or surgical scar in the acoustic path, and flow measurement is inaccurate when the ultrasound beam cannot be properly aligned with the axis of flow (Hoskins, 1996; Winkler & Wu, 1995). Two alternative approaches to 3D flow assessment are currently available to the researcher and clinician: (i) direct, model-independent velocity mapping using flow-encoded magnetic resonance imaging (MRI), and (ii) model-based computational fluid dynamics (CFD) calculations.

MRI is potentially the most appropriate technique for addressing all aspects of cardiovascular disease examination. MRI overcomes the acoustical window limitations of ultrasound, potentially allowing flow measurements to be obtained along any direction, and for any vessel in the cardiovascular system. MRI measurements are also less operator-dependent than those of Doppler ultrasound, and the true direction of flow can generally be precisely measured. The current gold standard for MRI flow quantification is phase contrast (PC) (O'Donnell, 1985). However, PC suffers from partial-volume effects when a wide distribution of velocities is contained within a single voxel (Tang et al., 1993). This is particularly problematic when flow is turbulent and/or complex (e.g., flow jets due to stenosis) or at the interface between blood and vessel wall (viscous sublayer). This issue is typically addressed by increasing the spatial resolution, which dramatically affects the signal-to-noise ratio (SNR) and increases the scan time. Therefore, PC may be inadequate for estimating the peak velocity of stenotic flow jets and for assessing wall shear rate.

An alternative approach for cardiovascular flow assessment is to reconstruct a complex flow field via CFD simulation, using vascular geometries and input/output functions derived from non-invasive imaging data (Kim et al., 2006; Papathanasopoulou et al., 2003; Steinman et al., 2002). The accuracy of conventional CFD routines hinges on many modeling assumptions that are not strictly true for *in vivo* vascular flow, including rigid vessel walls and uniform blood viscosity. Furthermore, conventional CFD routines typically employ a structured non-Cartesian finite-element mesh that conforms to the vessel geometry, which leads to relatively complex algorithms that incur significant computational costs.

This chapter proposes two new approaches for MRI assessment of carotid flow.

First, we introduce a rapid MRI method for fully-localized measurement of cardiovascular blood flow. The proposed method consists of combining slice-selective spiral imaging with Fourier velocity-encoded (FVE) imaging. The "spiral FVE" method provides intrinsically higher SNR than PC. Furthermore, it is not affected by partial-volume effects, as it measures the velocity distribution within each voxel. This makes this method particularly useful for assessment of flow in stenotic vessels. We show that spiral FVE measurements of carotid flow show good agreement with Doppler ultrasound measurements. We also introduce a mathematical model for deriving spiral FVE velocity distributions from PC velocity maps, and we use this model to show that spiral FVE provides good quantitative agreement with PC measurements, in an experiment with a pulsatile carotid flow phantom. Then, we show that it is possible to accurately estimate *in vivo* wall shear rate (WSR) and oscillatory shear index (OSI) at the carotid bifurcation using spiral FVE.

Finally, we propose a hybrid MRI/CFD approach that integrates low-resolution PC flow measurements directly into the CFD solver. The feasibility of this MRI-driven CFD approach is demonstrated in the carotid bifurcation of a healthy volunteer. We show that MRI-driven CFD has a regularizing effect on the flow fields obtained with MRI alone, and produces flow patterns that are in better agreement with direct MRI measurements than CFD alone. This approach may provide improved estimates of clinically significant blood flow patterns and derived hemodynamic parameters, such as wall shear stress (WSS).

2. Magnetic resonance flow imaging

2.1 Basic principles of MRI

MRI is a modality uniquely capable of imaging all aspects of cardiovascular disease, and is a potential "one-stop shop" for cardiovascular health assessment. MRI can generate cross-sectional images in any plane (including oblique planes), and can also measure blood flow. The image acquisition is based on using strong magnetic fields and non-ionizing radiation in the radio frequency range, which are harmless to the patient.

The main component of a MRI scanner is a strong magnetic field, called the B_0 field. This magnetic field is always on, even when the scanner is not being used. Typically, MR is used to image hydrogen nuclei, because of its abundance in the human body. Spinning charged particles (or "spins"), such as hydrogen nuclei, act like a tiny bar magnet, presenting a very small magnetic field, emanating from the south pole to the north pole. In normal conditions, each nucleus points to a random direction, resulting in a null net magnetization. However, in the presence of an external magnetic field (such as the B_0 field), they will line up with that field. However, they will not all line up in the same direction. Approximately half will point north, and half will point south. Slightly more than half of these spins (about one in a million) will point north, creating a small net magnetization M_0, which is strong enough to be detected. The net magnetization is proportional to the strength of the B_0 field, so MRI scanners with stronger magnetic fields (e.g., 3 Tesla) provide higher SNR.

Other important components of the scanner are the gradient coils. There are typically three gradient coils (G_x, G_y, and G_z), that produce an intentional perturbation in the B_0 field when turned on ("played"). This perturbation varies linearly along each spatial direction (x, y and z), such that no perturbation exists at the iso-center of the magnet when these gradients are used. In the presence of an external magnetic field, the spins rotate about the axis of that field. B_0 is (approximately) spatially uniform, so all spins initially rotate at the same frequency (the Larmor frequency), $\omega = \gamma B_0$, where γ is the gyromagnetic ratio (γ = 42.6 MHz/Tesla for hydrogen protons). However, when any of the gradients is played, the magnetic field becomes spatially varying, and so does the rotation frequency of the spins. Therefore, G_x, G_y, and G_z are used to frequency-encode (or phase-encode) spatial position along the x, y and z directions, respectively.

The final major component of the MR scanner is the radio-frequency (RF) coil. This is used to transmit an RF "excitation" pulse to the body, and also to receive the frequency-encoded signal from the "excited" portion of the body. In practice, independent coils may be used for transmission and reception. The RF pulse is typically modulated to the Larmor frequency. While B_0 is aligned with the z-axis (by definition), B_1, which is a very weak magnetic field associated with the RF pulse, is aligned with the x-axis (also by definition). When the RF pulse is played, some of the spins which are in resonance with the RF pulse (i.e., rotating at the RF pulse's frequency) will now begin to rotate around the x-axis (thus the name magnetic resonance). This tilts the net magnetization towards the x-y plane, and the net magnetization will now have a component in the x-y plane, M_{xy}.

The RF pulse is typically designed to have a somewhat rectangular profile in Fourier domain, centered at the modulation frequency (e.g., a modulated windowed sinc). This implies that the RF pulse in fact contains a certain range of frequencies, thus all spins rotating within that range become "excited", or tilted towards the x-axis. So, by playing gradient(s) of appropriate amplitude, and designing the RF pulse accordingly, one can excite only a thin slice of the body, which correspond to the region containing all spins that are in resonance with the RF pulse's range of frequencies. Excitation profiles other than "slices" may also be obtained (e.g., a pencil beam, or cylindrical excitation (Hu et al., 1993)), by designing an appropriate gradient/pulse combination.

When the RF pulse is turned off, M_{xy} begins to rotate (at the Larmor frequency) around the z-axis, as the net magnetization begins to realign with B_0. This rotating magnetization generates an oscillating signal, which can be detected by the receive coil. The frequency content of the received signal can be used to obtain spatial information about the excited portion of the body. In order to frequency-encode spatial information, gradients are also played during signal acquisition. These are called readout gradients. For imaging a slice perpendicular to the z-axis (an axial image), G_z is played during excitation (for slice-selection), and G_x and G_y are played during acquisition. These can be switched, for acquiring sagittal or coronal images, or all three gradients may be used during both excitation and acquisition to image oblique planes.

When the readout gradients are played, the acquired signal at a particular time instant corresponds to the sum of sinusoidal signals generated by spins located at different regions of the body, each rotating at different frequencies corresponding to their spatial locations. If an axial slice is being acquired, for example, the demodulated signal value is equivalent to a sample of the Fourier transform $M(k_x, k_y)$ of the cross-sectional image $m(x, y)$. In this case, by changing the amplitudes of G_x and G_y during acquisition, one may acquire different samples of $M(k_x, k_y)$. In fact, by playing G_x and/or G_y, one can move along the k_x-k_y plane (which is known in MRI as "k-space"), collecting samples of $M(k_x, k_y)$. When enough samples of $M(k_x, k_y)$ have been collected, an inverse Fourier transform produces $m(x, y)$.

The required coverage of k-space, and the number of samples, depend on the specified spatial resolution and field-of-view. For low spatial resolution imaging, only the central portion of k_x-k_y needs to be sampled. For higher spatial resolution, the periphery of k-space must also be covered. The field of view is associated with the spacing between samples. For a larger field-of-view, k-space needs to be more densely sampled, requiring an increased number of samples. If k-space is not sufficiently sampled, and the resulting field-of-view is not large enough to cover the entire object, overlap in spatial domain (aliasing) is observed.

Because signal amplitude is lost as the net magnetization realigns with B_0 (this is called relaxation), multiple acquisitions (excitation + readout) may be needed in order to cover the entire k-space. Some trajectories are more efficient in covering k-space than others. For example, spiral imaging, which uses oscillating gradients to achieve spiral k-space trajectories (Figure 1b), are generally faster than 2DFT imaging, i.e., require fewer acquisitions. In 2DFT imaging, each acquisition readout acquires a single line of k-space, sampling k_x-k_y in a Cartesian fashion (Figure 1a). This is generally slower, but may be advantageous in some applications with respect to the nature of associated image artifacts. The full sequence of RF pulses and gradients is called a "pulse sequence". The time between acquisitions is called the pulse repetition time, or TR.

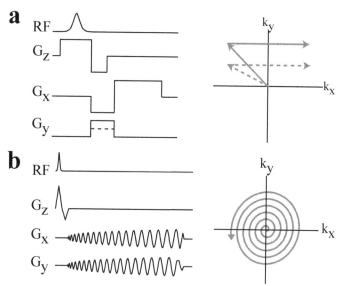

Fig. 1. Timing diagram (left) and corresponding k-space trajectories (right) for (a) 2DFT, and (b) spiral acquisitions.

2.2 Mathematical formalism

As discussed on section 2.1, the acquired MR signal $s(t)$ at a particular time instant corresponds to a sample of the Fourier transform $M(k_x, k_y)$ of the excited magnetization $m(x, y)$:

$$M(k_x, k_y) = \int_x \int_y m(x, y)\, e^{-j2\pi(k_x x + k_y y)}\, dx\, dy. \tag{1}$$

The Fourier coordinates k_x and k_y vary with time, according to the zeroth moment of the readout gradients G_x and G_y:

$$k_x(t) = \frac{\gamma}{2\pi} \int_0^t G_x(\tau)\, d\tau \tag{2}$$

$$k_y(t) = \frac{\gamma}{2\pi} \int_0^t G_y(\tau)\, d\tau. \tag{3}$$

These equations explain how the gradients can be used to "move" along k-space, as discussed in section 2.1. This formalism can be generalized for any combination of G_x, G_y, and G_z gradients:

$$M(\vec{k}_r) = \int_{\vec{r}} m(\vec{r}) \cdot e^{-j2\pi \vec{k}_r \cdot \vec{r}}\, d\vec{r} \tag{4}$$

$$\vec{k}_r(t) = \frac{\gamma}{2\pi} \int_0^t \vec{G}_r(\tau)\, d\tau, \tag{5}$$

where $\vec{G}_r$ is the oblique gradient resulting from the combination of the G_x, G_y and G_z gradients, and $\vec{r}$ is its corresponding axis along which the linear variation in magnetic field intensity is realized.

Given a spatial position function $\vec{r}(t)$ and a magnetic field gradient $\vec{G}_r(t)$, the magnetization phase is:

$$\phi(\vec{r}, t) = \gamma \int_0^t \vec{G}_r(\tau) \cdot \vec{r}(\tau)\, d\tau, \tag{6}$$

For static spins, $\vec{r}(t)$ is constant ($\vec{r}$), and this becomes:

$$\phi = \gamma \vec{r} \cdot \int_0^t \vec{G}_r(\tau)\, d\tau \tag{7}$$

$$= 2\pi \vec{k}_r \cdot \vec{r}, \tag{8}$$

as in the exponential in equation 4.

2.3 Principles of MR flow imaging

The basic principles of quantitative flow measurement using magnetic resonance were first proposed by Singer (1959) and Hahn (1960) in the late 1950s. However, clinical applications of MR flow quantification were not reported until the early 1980s (Moran et al., 1985; Nayler et al., 1986; Singer & Crooks, 1983; van Dijk, 1984). Current MR flow imaging methods are based on the fact that spins moving at a constant velocity accrue a phase proportional to the velocity times the first moment of the gradient waveform along the direction in which they are moving.

For spins moving along the $\vec{r}$-axis with a constant velocity $\vec{v}$, and initial position $\vec{r}_0$, we can write $\vec{r}(t) = \vec{r}_0 + \vec{v}t$. Rewriting equation 6, for $t = t_0$:

$$\phi = \gamma \int_0^{t_0} \vec{G}_r(t) \cdot (\vec{r}_0 + \vec{v}t)\, dt \tag{9}$$

$$= \gamma \vec{r}_0 \cdot \int_0^{t_0} \vec{G}_r(t)\, dt + \gamma \vec{v} \cdot \int_0^{t_0} \vec{G}_r(t)\, t\, dt \tag{10}$$

$$= \gamma \vec{r}_0 \cdot \vec{M}_0 + \gamma \vec{v} \cdot \vec{M}_1, \tag{11}$$

where $\vec{M}_0$ and $\vec{M}_1$ are the zeroth and first moments of the $\vec{r}$-gradient waveform at the time of signal acquisitions ("echo time", or "time to echo" (TE)), respectively. Thus, if a gradient with null zeroth moment is used (e.g., a bipolar gradient, aligned with $\vec{v}$), the phase accrued for a constant velocity spin is $\phi = \gamma \vec{v} \cdot \vec{M}_1$.

Therefore, if a bipolar gradient waveform is played between the excitation and the readout, the phase measured in a pixel of the acquired image is directly proportional to the velocity of the spins contained within its corresponding voxel. However, factors other than flow (such as inhomogeneities of the magnetic field) may cause additional phase shifts that would cause erroneous interpretation of the local velocity (Rebergen et al., 1993).

2.4 Phase contrast

The phase contrast method addresses the problem mentioned above by using two gradient-echo data acquisitions in which the first moment of the bipolar gradient waveform is varied between measurements (O'Donnell, 1985). The velocity in each voxel is measured as:

$$v(x,y) = \frac{\phi_a(x,y) - \phi_b(x,y)}{\gamma(M_1^a - M_1^b)}, \tag{12}$$

where $\phi_a(x,y)$ and $\phi_b(x,y)$ are the phase images acquired in each acquisition, and M_1^a and M_1^b are the first moment of the bipolar gradients used in each acquisition.

2.5 Fourier velocity encoding

While phase contrast provides a single velocity measurement associated with each voxel, Fourier velocity encoding (Moran, 1982) provides a velocity histogram for each spatial location, which is a measurement of the velocity distribution within each voxel.

FVE involves phase-encoding along a velocity dimension. Instead of only two acquisitions, as in phase contrast, multiple acquisitions are performed, and a bipolar gradient with a different amplitude (and first moment) is used in each acquisition. Equation 10 can be rewritten as:

$$\phi(\vec{r}, \vec{v}, t) = 2\pi(\vec{k}_r \cdot \vec{r} + \vec{k}_v \cdot \vec{v}), \tag{13}$$

where $\vec{k}_v$ is the velocity frequency variable associated with $\vec{v}$, and is proportional to the first moment of $\vec{G}_r(t)$:

$$\vec{k}_v = \frac{\gamma}{2\pi}\vec{M}_1. \tag{14}$$

Each voxel of the two-dimensional image is associated with a distribution of velocities. This three-dimensional function $m(x, y, v)$ is associated with a three-dimensional Fourier space $M(k_x, k_y, k_v)$. Thus, an extra dimension is added to k-space, and multiple acquisitions are required to cover the entire k_x-k_y-k_v space. In order to move along k_v, a bipolar gradient with the appropriate amplitude (and first moment) is played before the k_x-k_y readout gradients, in each acquisition. Placing the bipolar gradient along the z-axis will encode through-plane velocities. Placing the bipolar gradient along x or y will encode in-plane velocities. Oblique flow can be encoded using a combination of bipolar gradients along the x, y and z axes. Each acquisition along k_v is called a velocity encode. The number of required velocity encodes depends on the desired velocity resolution and velocity field-of-view (the maximum range of velocities measured without aliasing). For example, to obtain a 25 cm/s resolution over a 600 cm/s field-of-view, 24 velocity encodes are needed. The spatial-velocity distribution, $m(x, y, v)$, is obtained by inverse Fourier transforming the acquired data, $M(k_x, k_y, k_v)$. If cine imaging (Glover & Pelc, 1988) is used, measurements are also time resolved, resulting in a four-dimensional dataset: $m(x, y, v, t)$.

2.6 FVE signal model

2DFT phase contrast provides two two-dimensional functions, $m(x, y)$ and $v_o(x, y)$, the magnitude and velocity maps, respectively. If these maps are measured with sufficiently high spatial resolution, and flow is laminar, one can assume that each voxel contains only one velocity, and therefore the spatial-velocity distribution associated with the object is approximately:

$$s(x, y, v) = m(x, y) \times \delta(v - v_o(x, y)), \tag{15}$$

where $\delta(v)$ is the Dirac delta function.

In 2DFT FVE, k-space data is truncated to a rectangular cuboid in k_x-k_y-k_v space. The associated object domain spatial-velocity blurring can be modeled as a convolution of the true object distribution, $s(x, y, v)$, with sinc$(x/\Delta x)$, sinc$(y/\Delta y)$, and sinc$(v/\Delta v)$, where Δx and Δy are the spatial resolutions along the x and y axes, respectively, and Δv is the velocity resolution, as follows:

$$\hat{s}(x, y, v) = [m(x, y) \times \delta(v - v_o(x, y))] * \text{sinc}\left(\frac{x}{\Delta x}\right) * \text{sinc}\left(\frac{y}{\Delta y}\right) * \text{sinc}\left(\frac{v}{\Delta v}\right), \tag{16}$$

where $\hat{s}(x, y, v)$ is the measured object distribution, and $*$ denotes convolution. This is equivalent to:

$$\hat{s}(x, y, v) = \left[m(x, y) \times \text{sinc}\left(\frac{v - v_o(x, y)}{\Delta v}\right)\right] * \left[\text{sinc}\left(\frac{x}{\Delta x}\right) \times \text{sinc}\left(\frac{y}{\Delta y}\right)\right]. \tag{17}$$

This approach for deriving FVE data from high-resolution velocity maps can be used for many simulation purposes (Carvalho et al., 2010).

3. Slice-selective FVE with spiral readouts (spiral FVE)

Phase contrast imaging is fast, but has limitations associated with partial-volume effects. On the other hand, 2DFT FVE addresses those limitations, but requires long scan times. Thus, we propose the use of slice-selective FVE MRI with spiral acquisitions. The proposed spiral FVE method is capable of acquiring fully localized, time-resolved velocity distributions in a short breath-hold. Scan-plane prescription is performed using classic protocols.

We present practical implementations for measuring blood flow through the carotid arteries, and comparisons with Doppler ultrasound and high-resolution 2DFT phase contrast MRI. The proposed method is demonstrated in healthy volunteers. Subjects provided informed consent, and were imaged using a protocol approved by the institutional review board of the University of Southern California.

3.1 Pulse sequence

The spiral FVE imaging pulse sequence (Figure 2) consists of a slice-selective excitation, a velocity-encoding bipolar gradient, a spiral readout, and refocusing and spoiling gradients. The dataset corresponding to each temporal frame is a stack-of-spirals in k_x-k_y-k_v space (Figure 3). The bipolar gradient effectively phase-encodes in k_v, while each spiral readout acquires one "disc" in k_x-k_y.

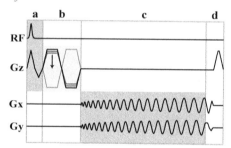

Fig. 2. Spiral FVE pulse sequence. It consists of (a) slice selective excitation, (b) velocity encoding bipolar gradient, (c) spiral readout, and (d) refocusing and spoiling gradients.

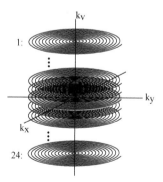

Fig. 3. Spiral FVE k-space sampling scheme. The dataset corresponding to each temporal frame is a stack-of-spirals in k_x-k_y-k_v space. Each spiral acquisition corresponds to a different k_v encode level.

3.2 Spiral FVE signal model

As spiral FVE acquisitions follow a stack-of-spirals pattern in k_x-k_y-k_v space (Figure 3), k-space data is truncated to a cylinder, i.e., a circle along k_x-k_y (with diameter $1/\Delta r$), and a rect function along k_v (with width $1/\Delta v$), where Δr and Δv are the prescribed spatial and velocity resolutions, respectively. Using the same approach we used in section 2.6 for 2DFT FVE, the associated object domain spatial-velocity blurring in spiral FVE can be modeled as a convolution of the true object distribution, $s(x, y, v)$, with $\mathrm{jinc}(\sqrt{x^2 + y^2}/\Delta r)$ and $\mathrm{sinc}(v/\Delta v)$, resulting in:

$$\hat{s}(x, y, v) = [m(x, y) \times \delta(\, v - v_0(x, y)\,)] * \mathrm{sinc}\left(\frac{v}{\Delta v}\right) * \mathrm{jinc}\left(\frac{\sqrt{x^2 + y^2}}{\Delta r}\right)$$

$$= \left[m(x, y) \times \mathrm{sinc}\left(\frac{v - v_0(x, y)}{\Delta v}\right)\right] * \mathrm{jinc}\left(\frac{\sqrt{x^2 + y^2}}{\Delta r}\right). \tag{18}$$

3.3 *In vitro* validation

An *in vitro* comparison of velocity distributions measured with spiral FVE with those derived from high-resolution 2DFT phase contrast — the current MR gold standard — was performed. The signal model presented in section 3.2 was used to generate simulated FVE data based on high-resolution 2DFT phase contrast data.

The validation experiments were performed using a pulsatile carotid flow phantom (Phantoms by Design, Inc., Bothell, WA). A slice perpendicular to the phantom's carotid bifurcation was prescribed, and through-plane velocities were measured. A cine gradient-echo 2DFT phase contrast sequence with high spatial resolution and high SNR (0.33 mm resolution, 10 averages, 80 cm/s Venc) was used as a reference. Cine spiral FVE data with $\Delta r = 3$ mm and $\Delta v = 10$ cm/s was obtained from the same scan plane. Both acquisitions were prospectively gated, and used the same TR (11.6 ms), flip angle (30°), slice profile (3 mm), temporal resolution (23.2 ms), and pre-scan settings. The total scan time was 40 minutes for phase contrast, and 12 seconds for FVE.

A simulated spiral FVE dataset was computed from the PC magnitude and velocity maps, using the convolution model described in Eq. 18. The PC-derived and FVE-measured data were registered by taking one magnitude image $m(x, y)$ from each dataset, and then using the phase difference between their Fourier transforms $M(k_x, k_y)$ to estimate the spatial shift between the images. Amplitude scaling was performed by normalizing the ℓ_2-norm of each FVE dataset. The difference between PC-derived and FVE-measured time-velocity distributions was calculated for select voxels, and the associated signal-to-error ratios were computed. This was used as a quantitative assessment of spiral FVE's accuracy.

Figure 4 shows measured and PC-derived time-velocity FVE distributions from nine representative voxels, selected around the circumference of the vessel wall of the pulsatile carotid flow phantom's bifurcation. The signal-to-error ratio between measured and PC-derived time-velocity distributions was measured to be within 9.3–11.7 dB. Imperfect registration between the datasets, combined with spatial blurring due to off-resonance in the measured spiral FVE data, may have contributed to this moderate signal-to-error ratio. Nevertheless, the two datasets show good visual agreement, and no significant spatial variation was observed in terms of accuracy. These results show that velocity distributions

measured with spiral FVE agree well with those obtained with 2DFT phase contrast, the
current MRI gold standard.

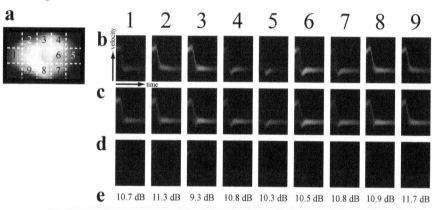

Fig. 4. *In vitro* evaluation of the accuracy of spiral FVE velocity histograms. Results are
shown for nine representative voxels, selected around the circumference of the vessel wall of
the pulsatile carotid flow phantom's bifurcation (a). For each voxel, it is shown: (b)
time-velocity distribution derived from high-resolution 2DFT phase contrast; (c)
time-velocity distribution measured with spiral FVE; (d) absolute difference between spiral
FVE and 2DFT PC-derived histograms; (e) signal-to-error ratio.

3.4 *In vivo* evaluation
The spiral FVE method was evaluated *in vivo*, aiming at quantifying flow through the common
carotid artery of a healthy volunteer. Doppler ultrasound was used as a gold standard and was
qualitatively compared with the proposed method. We also show experiments for evaluating
the most appropriate view-ordering scheme for measuring carotid flow, and we demonstrate
spiral FVE's ability to measure multiple flows in a single acquisition.

3.4.1 View ordering
Experiments were performed to determine the appropriate view-ordering scheme for carotid
spiral FVE studies. Flow was measured through the carotid artery of a healthy volunteer,
using a 4-interleaf spiral FVE acquisition. The measurement was performed twice, and
in each acquisition, a different view-ordering scheme was used. In the first acquisition,
one spiral interleaf was acquired per heartbeat, and two different k_v levels were encoded
throughout each R-R interval. In the second acquisition, two spiral interleaves were acquired
per heartbeat, encoding one k_v level per cardiac cycle. Reconstructed velocity histograms were
compared with respect to data inconsistency artifacts related to view-ordering.
The results are shown in Figure 5. Note that the ghosting artifacts that arise in the velocity
histograms when acquiring two different velocity encode levels during the same heartbeat
(Figure 5a) do not appear when we used interleaf segmentation instead (Figure 5b). In contrast
to view-sharing along k_v, which causes ghosting in the velocity direction, view-sharing
among spiral interleaves introduces swirling artifacts in image domain, reducing the effective
unaliased spatial field-of-view by a factor of 2. However, only moving spins (flowing blood)
experience these artifacts, and vessels on the same side of the neck are relatively close to each

other. Therefore, the unaliased field-of-view is wide enough to enclose all vessels on one side of the neck, so that quantification of a vessel of interest will not be disturbed by flow from neighboring vessels (e.g. measurement of flow in the left carotid arteries will not be disturbed by flow in the left jugular vein). In order to suppress signal from the opposite side of the neck, we separately reconstruct data from the left and right neck-coil elements. This uses the receiver coil sensitivity profile to avoid spiral view-sharing artifacts. Based on these observations, we used spiral interleaf view-sharing in all subsequent carotid studies.

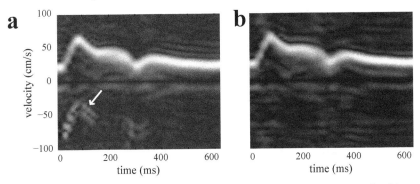

Fig. 5. Comparison of different view-orderings for multi-shot spiral FVE, in a healthy volunteer carotid study. When two or more k_v levels are acquired during the same heartbeat (a), velocity distribution changes between consecutive TRs cause ghosting artifacts along the velocity axis (arrow). This artifact is not seen if, in the same heartbeat, different spiral interleaves, but only one k_v encoding, are acquired (b).

3.4.2 Comparison with Doppler ultrasound

Representative *in vivo* results are compared with Doppler ultrasound in Figure 6. The MRI measured time-velocity histogram shows good agreement with the ultrasound measurement, as the peak velocity and the shape of the flow waveform were comparable to those observed in the ultrasound studies.

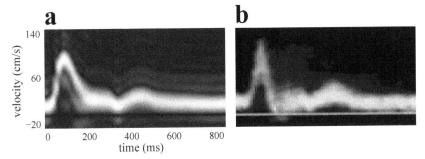

Fig. 6. Comparison of the spiral FVE method (a) with Doppler ultrasound (b), in a healthy volunteer common carotid artery study. Peak velocity and time-velocity waveforms are in good agreement.

3.4.3 Measurement of multiple flows

Figure 7 illustrates spiral FVE's ability to resolve different flows from a single dataset. A different time-velocity distribution was calculated for each voxel, and the distributions from select voxels are presented.

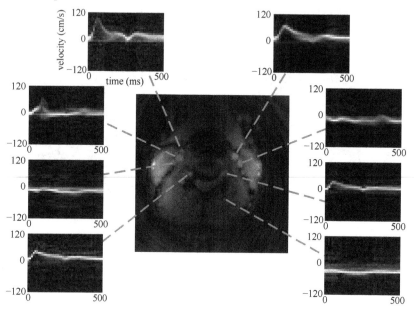

Fig. 7. Multiple flow distributions obtained from a single spiral FVE dataset. For each voxel in the image, a time-velocity distribution was calculated.

4. Wall shear rate estimation using spiral FVE

Arterial wall shear stress, the drag force acting on the endothelium as a result of blood flow, is widely believed to influence the formation and growth of atherosclerotic plaque, and may have prognostic value. According to Newton's law of viscosity, WSS can be estimated as the product of wall shear rate and blood viscosity (μ), where WSR is the radial gradient of blood flow velocity (dv/dr) at the vessel wall. Low WSS (Tang et al., 2008; Zarins et al., 1983) and highly oscillatory WSS (Ku et al., 1985) have been linked to the formation and growth of atherosclerotic plaques, and this link has been validated *in vitro* (Dai et al., 2004). High WSS has also been hypothesized as a factor responsible for the topography of atherosclerotic lesions (Thubrikar & Robicsek, 1995).

Phase contrast MRI (PC-MRI) suffers from data inconsistency, partial-volume effects (Tang et al., 1993), intravoxel phase dispersion and inadequate SNR at high spatial resolutions (Figure 8). PC-MRI is not currently capable of providing accurate absolute measurements of WSS (Boussel et al., 2009), and has been shown to underestimate blood flow velocities in the carotid bifurcation by 31–39% (Harloff et al., 2009).

As shown in section 3, spiral FVE produces accurate velocity histograms compared with those acquired with Doppler ultrasound. In addition, spiral FVE method is capable of rapid acquisition of fully localized, time-resolved velocity distributions. In this section, we propose

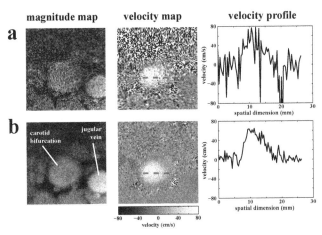

Fig. 8. Illustration of *in vivo* high-resolution 2DFT phase contrast MRI, obtained at the carotid bifurcation of a healthy volunteer at peak flow: (a) single acquisition; (b) ten signal averages. High spatial resolution is typically associated with low SNR (a). Averaging multiple acquisitions improves SNR (b), but also increases the total scan time, and may cause loss of effective resolution due to subject motion. Scan parameters: 0.33x0.33x3 mm^3 spatial resolution, 37 ms temporal resolution, 30° flip angle, 80 cm/s Venc, 2-minute scan (120 heartbeats) per acquisition.

the use of spiral FVE to estimate WSR and the oscillatory shear index. This is made possible by the method proposed by Frayne & Rutt (1995) for FVE-based WSR estimation, which is discussed next.

4.1 FVE-based WSR estimation: the Frayne method

The Frayne method for FVE-based WSR estimation (Frayne & Rutt, 1995) involves obtaining the velocity distribution for a voxel spanning the blood/vessel wall interface, and then using this distribution to reconstruct the velocity profile across the voxel, with sub-voxel spatial resolution. Assuming that signal intensity is spatially and velocity invariant, the sum of the signals from all velocities within a voxel is proportional to the total volume of material within the voxel. Furthermore, two assumptions can be made about the shape of the velocity profile: (i) the fluid velocity at the vessel wall is approximately zero, and (ii) the velocity profile within a voxel is monotonically increasing or decreasing. Using these assumptions, a step-wise discrete approximation $\tilde{v}(r)$ to the true velocity profile across a voxel can be obtained from its velocity distribution $s(v)$ by inverting the discrete function $r(v)$, which is constructed as follows:

$$r(v_i) = r(v_{i-1}) + \Delta r \cdot h(v_i), \quad \text{where} \quad h(v_i) = \frac{|s(v_i)|}{\sum_v |s(v)|}. \tag{19}$$

Note that, for each velocity bin v_i, the intra-voxel position r is incremented by a fraction of the total radial extent of the voxel (Δr). This fraction is proportional to $h(v_i)$, which is the volume fraction of the voxel that has velocity $v = v_i$. The volume fractions are calculated by normalizing the velocity distribution. This process is demonstrated graphically in Figure 9. Spatial variations in signal intensity due to radio-frequency saturation (i.e., inflow enhancement) and due to differences in ^1H density and relaxation properties between vessel wall tissue and blood may be compensated by adjusting $s(v)$ accordingly, prior to calculating $h(v)$ (Frayne & Rutt, 1995).

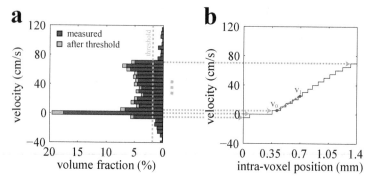

Fig. 9. Graphical illustration of the Frayne method (Frayne & Rutt, 1995). WSR is estimated from FVE velocity distributions in voxels spanning the blood/vessel wall interface. First, a threshold is applied to the velocity histogram to reduce noise sensitivity (a). Then, the volume fraction within each velocity bin is converted into a radial position across the voxel, using Eq. 19. Finally, the velocity gradient is estimated from the reconstructed velocity profile (b), within a small velocity interval $[v_0, v_1]$.

In order to reduce the effects of ringing and noise rectification due to the magnitude operation in Eq. 19, a threshold is applied to $s(v)$ before normalization (Figure 9a). The appropriate threshold level must be determined by analyzing the signal intensities in the velocity distribution for a range of velocities outside the range of expected blood flow velocities. It is assumed that signal outside this expected range is exclusively due to rectified noise (Frayne & Rutt, 1995). In our implementation, only components that are below the specified threshold *and* outside this expected range of velocities are set to zero.

The WSR can be estimated by prescribing a velocity interval $[v_0, v_1]$ and then fitting a first-order polynomial to the points of $\tilde{v}(r)$ within this interval (Figure 9b). Ideally, $v_0 = 0$ and $v_1 = \Delta v$, because we wish to estimate the velocity derivative at the blood-wall interface. The SNR of shear rate estimates will increase as this velocity interval becomes larger, because of averaging across multiple velocity steps. However, as the interval becomes larger, the shear rate is averaged over a larger distance within the voxel and may deviate from the true local shear at the wall (Frayne & Rutt, 1995). Therefore, it is important to prescribe a reasonable $[v_0, v_1]$ interval. In our implementation, $v_0 = \Delta v$ and $v_1 \approx 30$ cm/s are used for an initial assessment, and then the interval is manually adjusted for selected voxels of interest. The same voxel-based approach is used with respect to the noise threshold discussed above, with a fixed threshold value being used for the initial assessment. A negative $[v_0, v_1]$ interval — e.g., $v_0 = -5$ cm/s and $v_1 = -15$ cm/s — is used when large volume fractions are measured on negative velocities, i.e., when there are large $h(v)$ values for $v < 0$. This allows measurements of negative WSR values.

4.2 *In vivo* WSR measurement

The *in vivo* measurement of carotid WSR using spiral FVE acquisitions with Frayne's reconstruction is now demonstrated. Three healthy subjects were studied. For each volunteer, five 5 mm contiguous slices (2.5 cm coverage) were prescribed perpendicular to the left carotid bifurcation. Each slice was imaged independently (separate acquisitions). Localized gradient shimming was performed, and acquisitions were prospectively ECG-gated. Several cardiac phases were acquired, spanning the systolic portion of the cardiac cycle. A time-bandwidth product 2 RF pulse was used for excitation, and the flip angle was 30°. Only through-plane velocities were measured, using 32 k_v encoding steps over a 160 cm/s velocity field-of-view (5 cm/s resolution). The velocity field-of-view was shifted from the -80 to 80 cm/s range to the -40 to 120 cm/s range during reconstruction, in order to avoid aliasing and maximize field-of-view usage. Negative velocities are encoded in order to assess negative WSR values, and also to accommodate leakage and ringing due to k_v truncation (i.e., finite velocity resolution). Spatial encoding was performed using eight 4 ms variable-density spiral interleaves (16~4 cm field-of-view, 1.4 mm resolution). The temporal resolution was 24 ms (12 ms TR, 2 views per beat). Scan time was 128 heartbeats per slice, i.e., approximately 2 minutes per slice. The subjects provided informed consent, and were scanned using a protocol approved by the institutional review board of the University of Southern California. Figures 10 and 11 show two representative sets of *in vivo* results. The WSR values are shown for manually-segmented regions-of-interest. Figure 10 illustrates the variation in WSR along all three spatial dimensions near the carotid bifurcation of subject #1. These results correspond to the cardiac phase with the highest peak velocity. Figure 11 illustrates the temporal variation of pulsatile WSR in the common carotid artery of subject #2. The results show a circumferential variation in WSR around the wall of the common carotid artery. Markl et al. (2009) recently observed a similar variation using a PC-based approach.

4.3 Estimation of oscillatory shear index

The OSI is important for the evaluation of shear stress imposed by pulsatile flow. This index describes the shear stress acting in directions other than the direction of the temporal mean shear stress vector (He & Ku, 1996; Ku et al., 1985). It is defined as the relation between the time-integral of the shear stress component acting in the direction opposite to the main direction of flow and the time-integral of the absolute shear stress. In this work, OSI was calculated as defined by He & Ku (1996), and assuming spatially and temporally invariant blood viscosity.

Measuring negative WSR with the proposed method requires voxels to be small enough to contain only reverse flow. Voxels containing both forward and reverse flow would violate the assumption of a monotonically increasing/decreasing velocity profile within the voxel. Under sufficient spatial resolution, negative WSR may be measured simply by using a negative $[v_0, v_1]$ interval.

Figure 12 presents a demonstration of *in vivo* OSI estimation, using the proposed method. Results are shown for eight representative voxels, selected around the circumference of the wall of the carotid bifurcation of subject #3. Measured velocity distributions, wall shear rates, and OSI values, corresponding to the systolic portion of the cardiac cycle, are shown for each voxel. High OSI values were observed at the wall corresponding to the carotid bulb. Non-zero OSI was also observed at the opposite wall. These findings are in agreement with PC-based OSI measurements recently reported by Markl et al. (2009). The results also suggest

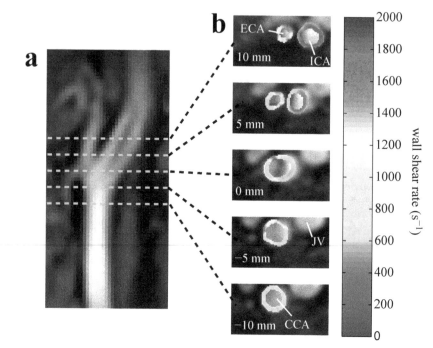

Fig. 10. Carotid WSR measured across the carotid bifurcation of subject #1: (a) slice prescription; (b) spiral FVE WSR measurements. Results correspond to the cardiac phase with the highest peak velocity (96 ms after the ECG trigger). The common (CCA), external (ECA), and internal (ICA) carotid arteries, and the jugular vein (JV), are indicated.

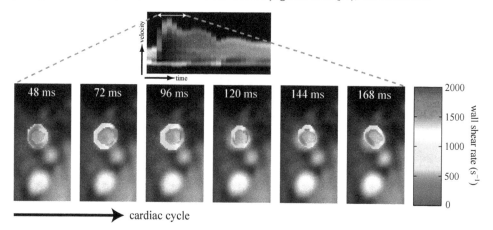

Fig. 11. Temporal variation of pulsatile WSR measured in the common carotid artery of subject #2. Results correspond to the cardiac phases acquired 48–168 ms after the ECG trigger, and to a slice prescribed 10 mm below the carotid bifurcation.

that higher spatial resolution is needed for accurately estimating OSI in some of the voxels. Notably, voxel (h) presents both positive and negative velocity components simultaneously during post-systolic deceleration. This indicates that the spatial resolution was insufficient, and the assumption of a monotonically decreasing velocity profile within the voxel was violated.

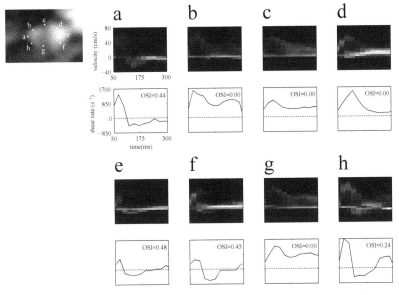

Fig. 12. *In vivo* assessment of oscillatory shear index. Results are shown for eight representative voxels (a–h), selected around the circumference of the wall of the carotid bifurcation of subject #3 (see inset). Measured velocity distributions, wall shear rates, and OSI values, corresponding to the systolic portion of the cardiac cycle, are shown for each voxel.

5. Assessment of carotid flow using CFD (and MRI)

Computational fluid dynamics methods are concerned with the approximate solution of the fluid motion as well as with the interaction of the fluid with solid bodies. Fluid dynamic problems, essentially, are based on nonlinear systems of coupled partial differential equations. Attempting to solve those systems analytically, in general, is an impracticable task. Over the years, researchers have developed methodologies, codes, schemes and algorithms to find approximate solutions, focusing on accuracy and speed of convergence, for problems involving fluid dynamics and heat transfer.

Historically, CFD started in the early 1970s, triggered by the availability of increasingly more powerful mainframes. In the beginning, the study was limited to high-technology engineering areas of aeronautics and astronautics. Nowadays, computational fluid dynamics methodologies are routinely employed in many fields such as: racing car design, ship design, meteorology, oil recovery, civil engineering, airspace engineering, and biomedical engineering.

This section is dedicated to the use of CFD for evaluation of blood flow in the carotid arteries. *In vivo* 3D blood flow patterns can be either measured directly using PC-MRI, or obtained from model-based CFD calculations. PC-MRI is accurate, but suffers from long scan times and limited spatio-temporal resolution and SNR. CFD provides arbitrarily high resolution and reduced scan times, but its accuracy hinges on the model assumptions. A numerical framework for constructing a flow field that is influenced by both direct measurements and a fluid physics model is described.

5.1 CFD in biomedical engineering

Recently, CFD is playing an important role in the analysis of blood flow. This technique of flow visualization has been widely applied in problems involving arterial diseases with reconstruction of hemodynamics in realistic models based on images generated by standard *in vivo* medical visualization tools, such as MRI. CFD can be used to improve the data obtained using MRI for estimation of the flow properties of blood vessels. CFD can also be used to compare real data obtained from patients with simulated data models using realistic geometries. These geometries may even be constructed using MRI data.

Boussel et al. (2009) compared a time-dependent 3D phase-contrast MRI sequence with patient-specific CFD models for patients who had intracranial aneurysms. The evolution of intracranial aneurysms is known to be related to hemodynamic forces such as wall shear stress and maximum shear stress. Harloff et al. (2010) used CFD image-based modeling as an option to improve the accuracy of MRI-based WSS and OSI estimation, with the purpose of correlating atherogenic low WSS and high OSI with the localization of aortic plaques. Steinman et al. (2002) used a novel approach for noninvasively reconstructing artery wall thickness and local hemodynamics at the human carotid bifurcation. Three-dimensional models of the lumen and wall boundaries, from which wall thickness can be measured, were reconstructed from black blood MRI. Along with time-varying inlet/outlet flow rates measured via PC-MRI, the lumen boundary was used as input for a CFD simulation of the subject-specific flow patterns and wall shear stress. Long et al. (2003) were concerned with the reproducibility of geometry reconstruction, one of the most crucial steps in the modeling process. Canstein et al. (2008) used rapid prototyping to transform aortic geometries as measured by contrast-enhanced MR angiography into realistic vascular models with large anatomical coverage. Visualization of characteristic 3D flow patterns and quantitative comparisons of the *in vitro* experiments with *in vivo* data and CFD simulations in identical vascular geometries were performed to evaluate the accuracy of vascular model systems.

CFD can also be used with other imaging techniques such as tomography. Howell et al. (2007) used CFD to study the temporal and spatial variations in surface pressure and shear through the cardiac cycle on models of bifurcated stent-grafts derived from computed tomography in patients who had previously undergone endovascular repair of abdominal aortic aneurysm.

5.2 MRI-driven CFD

A numerical framework for constructing a flow field that is influenced by both direct measurements and a fluid physics model is now described. The PC-MRI signal is expressed as a linear function of the velocities on the underlying computational grid, and a tunable parameter controls the relative influence of direct measurements and the model assumptions. The feasibility of the proposed approach is demonstrated in the carotid bifurcation of a healthy volunteer. The results show that this methodology produces flow fields that are in better agreement with direct PC-MRI measurements than CFD alone.

5.2.1 Proposed approach

Here, blood is modeled as an incompressible Newtonian fluid with constant viscosity μ. This model is widely used in *in vivo* CFD analysis, and underlies the commonly held definition of wall shear stress as being proportional to dv/dr, where v is the velocity tangential to the vessel wall, and r is the perpendicular distance from the wall. The task of a CFD routine is to solve the momentum and continuity equations that the flow field — subject to the assumption of Newtonian flow — must obey. Following the control-volume formulation introduced by Patankar (1980), the momentum equation is

$$\rho \frac{D\mathbf{v}}{Dt} = -\nabla p + \mu \Delta \mathbf{v}, \tag{20}$$

where ρ is the fluid density, $\mathbf{v} = (u, v, w)$ is the velocity vector field, p is the pressure field, ∇ is the gradient operator, and Δ is the Laplacian operator. The flow field must also satisfy mass conservation (or continuity), which can be expressed as

$$\nabla \cdot \mathbf{v} = 0. \tag{21}$$

Equations (20) and (21) must be solved for the unknown scalar field variables u, v, w, and p. These equations are non-linear and coupled, and attempting to solve them directly in one step is a formidable, if not impossible, task.

The solver was built on the SIMPLER algorithm developed by Patankar (1980), which is a well-known and established numerical routine for solving Eqs. (20) and (21). SIMPLER starts with an initial estimate for the velocity field, and updates this estimate in an iterative fashion. At each iteration i, the discretized equations are linearized using the velocity estimate $\mathbf{v}_{i-1}$ at the previous iteration, which produces a square system matrix $\mathbf{A}$ for each velocity component. For example, for the z velocity component, we can write

$$\mathbf{A}_{i-1}\mathbf{w}_i = \mathbf{b}_{i-1}, \tag{22}$$

where $\mathbf{w}$ is the z velocity in all voxels in the 3D calculation domain (stacked to form a 1D column vector), and $\mathbf{b}$ is a constant column vector. Note that the pressure field is updated periodically based on the current velocity field estimate, and hence does not appear explicitly in (22).

The key step in our approach is to add additional rows to $\mathbf{A}$ and $\mathbf{b}$ that incorporate MRI measurements of one or more velocity components. The underlying assumption we make here is that the velocity measured with PC-MRI is equal to the average velocity within the voxel, and can hence be expressed as a linear combination of the velocities on the underlying CFD calculation grid. For example, for the z (S/I) velocity component we have

$$\mathbf{w}_{MR} \approx \mathbf{A}_{MR}\mathbf{w}, \tag{23}$$

where $\mathbf{w}_{MR}$ is the z velocity component measured with PC-MRI.

Combining Eqs. (22) and (23), we have

$$\begin{bmatrix} \mathbf{A}_{i-1} \\ s\mathbf{A}_{MR} \end{bmatrix} \mathbf{w}_i = \begin{bmatrix} \mathbf{b}_{i-1} \\ s\mathbf{w}_{MR} \end{bmatrix}. \tag{24}$$

The weighting parameter s is an adjustable scalar that determines the influence of the MRI measurements on the solution. For example, $s = 0$ produces a conventional, unconstrained CFD solution. Here, we solve Eq. (24) in the least squares sense, using the conjugate gradient method on the normal equations. Hence, the solution $\mathbf{w}$ at each iteration represents a weighted least squares estimate, with relative contributions of fluid physics and direct MRI measurement being controlled by the parameters after multiple interactions. The velocity field $\mathbf{v} = (u, v, w)$ converges toward the solution for a particular point t. Time is then incremented by an amount Δt, and the iterative procedure is repeated to obtain the solution at $t + \Delta t$. This way, it is possible to predict a time-dependent flow field, as long as the true velocity field $\mathbf{v}$ at $t = 0$ is known. In this work, however, the exact velocity field is not known, since only a relatively low-resolution and noisy PC-MRI velocity field is available. Instead, we will calculate the steady flow $\mathbf{v}_s$ subject to the measured inlet and outlet velocities. We will furthermore assume that the velocity field obtained with PC-MRI at one time-point (near peak flow) is representative of the steady flow given the inlet and outlet velocities at that time-point. We obtain $\mathbf{v}_s$ by starting with an initial guess for $\mathbf{v}$, and carrying the simulations forward in time until convergence.

5.2.2 *In vivo* demonstration

PC-MRI data were obtained from four time-resolved 3DFT FGRE image volumes in the carotid artery in one healthy volunteer ($1 \times 1 \times 2.5$ mm^3 voxel size; FOV $16 \times 12 \times 7.5$ cm^3; TR 7.0 ms; flip angle $15°$; temporal resolution 56 ms; Venc 1.6 m/s; 7 minutes per scan), on a GE Signa 3T EXCITE HD system (4 G/cm and 15 G/cm/ms maximum gradient amplitude and slew rate) with a 4-channel carotid receive coil array. The through-slab (z) axis was oriented along the S/I direction. The subject provided informed consent, and was scanned using a protocol approved by the institutional review board of the University of Southern California. PC-MRI velocity component maps u_{MR}, v_{MR}, and w_{MR} were calculated using data from one receive coil. Residual linear velocity offsets in each velocity component map (e.g. due to eddy-currents) were removed by performing a linear fit to manually defined 3D regions containing only stationary tissue. The vessel lumen was segmented by manually outlining the vessel borders from a stack of 2D axial slices.

The solver calculations assumed a blood viscosity of 0.0027 Pa·sec and a blood density of 1060 kg/m^3 (Reynolds number of order 1000), and no-slip boundary conditions. Calculations were performed on a Cartesian grid of 1 mm isotropic resolution, and all algorithms were implemented in Matlab. Low-resolution (truncated to $1 \times 1 \times 3$ mm^3) measurements of the S/I velocity component w_{MR} near the time-point of peak flow were incorporated into the solver, and a steady velocity vector field $\mathbf{v}_s$ was calculated as described above. Hence, $\mathbf{w}$ was partially constrained by the measured velocity component w_{MR}, whereas u and v were determined solely from the fluid physics model. As in the original SIMPLER algorithm (Patankar, 1980), u, v, and w were defined on regular grids that were staggered by half a grid spacing (in different directions) with respect to the centered grid. This is done to avoid a checkerboard solution for the pressure and velocity fields (Patankar, 1980). The calculation domain was rectangular of size $20.5 \times 16.5 \times 25$ mm^3.

Figure 13 compares flow fields in the carotid artery obtained with PC-MRI only, CFD only ($s = 0$), and the proposed combined solver algorithm with $s = 5$. These flow fields were calculated from 4, 1, or 2 time-resolved 3D image acquisitions, corresponding to a total scan time of 28, 7, and 14 minutes, as indicated in the figure. In the common carotid artery, flow is predominantly along z, and all methods produce comparable flow fields. In the bifurcation, the bulk flow pattern appears qualitatively similar for all three methods, which indicates that the underlying fluid physics model makes reasonable predictions regarding the transverse velocities. However, comparison of the CFD and PC-MRI results shows that CFD underestimates the velocities in the external carotid artery. The combined solver, which strikes a compromise between CFD and PC-MRI, brings the calculated flow field in closer agreement with the values measured with PC-MRI.

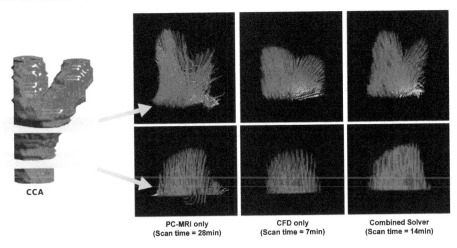

PC-MRI only (Scan time = 28min)	**CFD only** (Scan time = 7min)	**Combined Solver** (Scan time = 14min)

Fig. 13. Blood flow in the carotid bifurcation (top) and the common carotid artery (bottom) obtained with 4-point phase contrast MRI (left), CFD (center), and the combined solver ($s = 5$, right). The lines show the path of massless particles during the course of 60 ms (top) or 20 ms (bottom), under the assumption of a constant velocity field. Pathlines are RGB color-coded according to the local velocity direction (vertical=blue; in-plane=red and green). These velocity fields were obtained from 4 (left), 1 (center), or 2 (right) MRI scans, corresponding to total scan times of 28, 7 and 14 minutes, respectively. Compared to CFD, the combined solver produces flow fields that are in better qualitative and quantitative agreement with PC-MRI.

6. Conclusion

In this chapter, we have introduced spiral FVE, a rapid MRI method for fully-localized measurement of cardiovascular blood flow. The proposed method was shown to be capable of measuring blood flow in the carotid arteries and estimating wall shear stress and oscillatory shear index at the carotid bifurcation. In addition, we proposed a combined CFD-MRI solver that integrates the non-linear coupled system of the fluid partial differential equations using MRI data as initial data. This methodology, which has a tunable parameter, is capable of producing flow fields that are in better agreement with direct PC-MRI measurements than CFD alone.

MRI is potentially the most appropriate technique for addressing all aspects of a complete cardiovascular disease examination. The evaluation of carotid flow will be a necessary capability in such examination. The methodologies presented in this chapter can improve the quality of the MRI measured data, reducing scan time, and improving the SNR. These techniques should improve the diagnosis and understanding of carotid diseases.

7. References

Boussel, L., Rayz, V., Martin, A., Acevedo-Bolton, G., Lawton, M. T., Higashida, R., Smith, W. S., Young, W. L. & Saloner, D. (2009). Phase-contrast magnetic resonance imaging measurements in intracranial aneurysms in vivo of flow patterns, velocity fields, and wall shear stress: comparison with computational fluid dynamics, *Magn Reson Med* 61(2): 409–417.

Canstein, C., Cachot, P., Faust, A., Stalder, A. F., Bock, J., Frydrychowicz, A., Kuffer, J., Hennig, J. & Markl, M. (2008). 3D MR flow analysis in realistic rapid-prototyping model systems of the thoracic aorta: comparison with in vivo data and computational fluid dynamics in identical vessel geometries, *Magn Reson Med* 59(3): 535–546.

Carvalho, J. L. A., Nielsen, J. F. & Nayak, K. S. (2010). Feasibility of in vivo measurement of carotid wall shear rate using spiral Fourier velocity encoded MRI, *Magn Reson Med* 63(6): 1537–1547.

Dai, G., Kaazempur-Mofrad, M. R., Natarajan, S., Zhang, Y., Vaughn, S., Blackman, B. R., Kamm, R. D., Garcia-Cardena, G. & Gimbrone Jr., M. A. (2004). Distinct endothelial phenotypes evoked by arterial waveforms derived from atherosclerosis-susceptible and -resistant regions of human vasculature, *Proceedings of the National Academy of Sciences of the United States of America* 101(41): 14871–14876.

Frayne, R. & Rutt, B. K. (1995). Measurement of fluid-shear rate by Fourier-encoded velocity imaging, *Magn Reson Med* 34(3): 378–387.

Glover, G. H. & Pelc, N. J. (1988). A rapid-gated cine MRI technique, *Magn Reson Annu* pp. 299–333.

Hahn, E. L. (1960). Detection of sea-water motion by nuclear precession, *J Geophys Res* 65(2): 776–777.

Harloff, A., Albrecht, F., Spreer, J., Stalder, A. F., Bock, J., Frydrychowicz, A., Schollhorn, J., Hetzel, A., Schumacher, M., Hennig, J. & Markl, M. (2009). 3D blood flow characteristics in the carotid artery bifurcation assessed by flow-sensitive 4D MRI at 3T, *Magn Reson Med* 61(1): 65–74.

Harloff, A., Nussbaumer, A., Bauer, S., Stalder, A. F., Frydrychowicz, A., Weiller, C., Hennig, J. & Markl, M. (2010). In vivo assessment of wall shear stress in the atherosclerotic aorta using flow-sensitive 4D MRI, *Magn Reson Med* 63(6): 1529–1536.

He, X. & Ku, D. N. (1996). Pulsatile flow in the human left coronary artery bifurcation: average conditions, *J Biomech Eng* 118(1): 74–82.

Hoskins, P. R. (1996). Accuracy of maximum velocity estimates made using Doppler ultrasound systems, *Br J Radiol* 69(818): 172–177.

Howell, B. A., Kim, T., Cheer, A., Dwyer, H., Saloner, D. & Chuter, T. A. M. (2007). Computational fluid dynamics within bifurcated abdominal aortic stent-grafts, *J Endovasc Ther* 14(2): 138–143.

Hu, B. S., Pauly, J. M. & Macovski, A. (1993). Localized real-time velocity spectra determination, *Magn Reson Med* 30(3): 393–398.

Kim, C. S., Kiris, C., Kwak, D. & David, T. (2006). Numerical simulation of local blood flow in the carotid and cerebral arteries under altered gravity, *J Biomech Eng* 128(2): 194–202.

Ku, D. N., Giddens, D. P., Zarins, C. K. & Glagov, S. (1985). Pulsatile flow and atherosclerosis in the human carotid bifurcation. Positive correlation between plaque location and low oscillating shear stress, *Arterioscler Thromb Vasc Biol* 5: 293–302.

Long, Q., Ariff, B., Zhao, S. Z., Thom, S. A., Hughes, A. D. & Xu, X. Y. (2003). Reproducibility study of 3D geometrical reconstruction of the human carotid bifurcation from magnetic resonance images, *Magn Reson Med* 49(4): 665–674.

Markl, M., Wegent, F., Bauer, S., Stalder, A. F., Frydrychowicz, A., Weiller, C., Schumacher, M. & Harloff, A. (2009). Three-dimensional assessment of wall shear stress distribution in the carotid bifurcation, *Proc, ISMRM, 17th Annual Meeting*, Honolulu, p. 323.

Moran, P. R. (1982). A flow velocity zeugmatographic interlace for NMR imaging in humans, *Magn Reson Imaging* 1(4): 197–203.

Moran, P. R., Moran, R. A. & Karstaedt, N. (1985). Verification and evaluation of internal flow and motion. true magnetic resonance imaging by the phase gradient modulation method, *Radiology* 154(2): 433–441.

Nayler, G. L., Firmin, D. N. & Longmore, D. B. (1986). Blood flow imaging by cine magnetic resonance, *J Comput Assist Tomogr* 10(5): 715–722.

O'Donnell, M. (1985). NMR blood flow imaging using multiecho, phase contrast sequences, *Med Phys* 12(1): 59–64.

Papathanasopoulou, P., Zhao, S., Köhler, U., Robertson, M. B., Long, Q., Hoskins, P., Xu, X. Y. & Marshall, I. (2003). MRI measurement of time-resolved wall shear stress vectors in a carotid bifurcation model, and comparison with CFD predictions, *J Magn Reson Imaging* 17(2): 153–162.

Patankar, S. V. (1980). *Numerical Heat Transfer and Fluid Flow*, Hemisphere Publishing Corporation.

Rebergen, S. A., van der Wall, E. E., Doornbos, J. & de Roos, A. (1993). Magnetic resonance measurement of velocity and flow: technique, validation, and cardiovascular applications, *Am Heart J* 126(6): 1439–1456.

Singer, J. R. (1959). Blood flow rates by nuclear magnetic resonance measurements, *Science* 130(3389): 1652–1653.

Singer, J. R. & Crooks, L. E. (1983). Nuclear magnetic resonance blood flow measurements in the human brain, *Science* 221(4611): 654–656.

Steinman, D. A., Thomas, J. B., Ladak, H. M., Milner, J. S., Rutt, B. K. & Spence, J. D. (2002). Reconstruction of carotid bifurcation hemodynamics and wall thickness using computational fluid dynamics and MRI, *Magn Reson Med* 47(1): 149–159.

Tang, C., Blatter, D. D. & Parker, D. L. (1993). Accuracy of phase-contrast flow measurements in the presence of partial-volume effects, *J Magn Reson Imaging* 3(2): 377–385.

Tang, D., Yang, C., Mondal, S., Liu, F., Canton, G., Hatsukami, T. & Yuan, C. (2008). A negative correlation between human carotid atherosclerotic plaque progression and plaque wall stress: in vivo MRI-based 2D/3D FSI models, *J Biomech* 41(4): 727–736.

Thubrikar, M. J. & Robicsek, F. (1995).   Pressure-induced arterial wall stress and atherosclerosis, *Ann Thorac Surg* 59(6): 1594–1603.

van Dijk, P. (1984). Direct cardiac NMR imaging of heart wall and blood flow velocity, *J Comput Assist Tomogr* 8(3): 429–436.

Winkler, A. J. & Wu, J. (1995). Correction of intrinsic spectral broadening errors in Doppler peak velocity measurements made with phased sector and linear array transducers, *Ultrasound Med Biol* 21(8): 1029–1035.

Zarins, C. K., Giddens, D. P., Bharadvaj, B. K., Sottiurai, V. S., Mabon, R. F. & Glagov, S. (1983). Carotid bifurcation atherosclerosis. Quantitative correlation of plaque localization with flow velocity profiles and wall shear stress, *Circulation Research* 53(4): 502–514.

6

Biomechanical Factors Analysis in Aneurysm

Kleiber Bessa[1], Daniel Legendre[2] and Akash Prakasan[2]
[1]*Department of Environmental Sciences and Technological*
Rural Federal University of Semi-Arid
[2]*Institute Dante Pazzanese of Cardiology*
Brazil

1. Introduction

All the cells in the body need to receive food (nutrients, metabolic products) and to dispose of waste products. The responsible system for that is cardiovascular system. It is responsible to supply food through the arteries and to return waste products through the veins for all living cells in the human body. This task is reached by a circulating fluid, the blood. The central location which all lines of supply originate from and return to is a small, very small, pump, the heart. The heart keeps the fluid in circulation. In the heart, there are two pumps, propelling blood into the pulmonary and systemic circulation and are combined into a single muscular organ to synchronously beat. Any disruption in the blood flow causes a disruption in food supply. Life is not possible without blood, but in the truth life is not possible without the circulation of blood. It must pump at all times, which it does by contracting and relaxing in a rhythmic pattern, approximately once every second, more than 86 thousand times every day, and about 2 billion times in a lifetime of 75 years, nonstop (Zamir, 2005). The blood ejected by the heart follows in the direction the arterial tree. Along the arterial tree, the arteries successively decrease in size, increase in number, undergo structural changes, and finish in arterioles that are as little as 10 µm in diameter. The structure of the artery is quite complex. The main components of the vessel wall are endothelium, smooth muscle cells, elastic tissue, collagen, and connective tissue. The arteries are targets for diseases such as atherosclerosis or aneurysms that each year claims the lives of scores of people worldwide. The cardiovascular disease may be triggered or aggravated by mechanical stimuli, such as wall stress or stretch resulting from the blood pressure, or shear stress resulting from the blood flow (Wernig and Xu, 2002). Arteries can also adapt to long-term physiological conditions by thinning or thickening the muscular layer, and altering the relative composition and organization of the various assemblies of structural proteins in process generally know as remodelling. Bessa et al. (2011) showed that occurs remodelling in tail arterial bed from normotensive and hypertensive rats. As shown in Figure 1, the internal diameter of the proximal portion of the tail artery did not differ between Wistar rats and spontaneously hypertensive rats (SHR), whereas the diameter of the intermediate and distal portions of SHR tails arteries were significant smaller than those of normotensive rats.

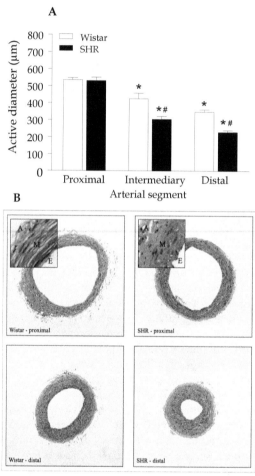

Fig. 1. *A*, Active internal diameters of tail arteries from Wistar rats (N=4animals) and spontaneously hypertensive rats (SHR, N=5 animals) fixed at a volumetric flow rate of 2.5 mL/min. *P <0.05 *vs* proximal portion. #P < 0.05 *vs* wistar rats (two-way ANOVA followed by the Bonferroni *post hoc* test). B, Photomicrograph of transverse section of proximal and distal tail artery from Wistar rats and SHR. Magnification: 10X. Inset panels in the proximal tail artery photomicrographs of Wistar and SHR are amplified images showing the endothelial (E), medium (M) and adventitial (A) layers of the artery. Extracted from Bessa et al. 2011.

An aneurysm is defined as a focal dilation of a blood vessel when compared with the original artery. Aneurysms are widening of the lumen in any artery, most commonly in the aorta for fusiform types, and in the head for saccular types. The arterial fusiform aneurysms and intrinsic stenosis are possible complications of atheroma. Saccular aneurysms can occur as complications of arterial wall trauma or infection. Intracranial aneurysms (IA), rare in childhood and adolescence, are observed in 3% to 5% of the population, with a gender ratio of 3 women for 2 men. Abdominal aortic aneurysms (AAA) most of the affect men between 40 and 70 years old (5-7% of people older than 60) (Thiriet, 2008). Genetics and risk factors play important roles in the development of the aneurysms, but is universally accepted that biomechanical factors (including increased pressures in hypertension) also play fundamentals roles. Vascular endothelial and smooth muscle cell are constantly exposed to the biomechanical factors caused by the blood flow. Cellular responses to these biomechanical factors influence vessel wall homeostasis (Hisai, 2008). Once the aneurysm forms, the biomechanical factors caused by the pulsatile flow in the aneurysm can cause gradual expansion of it. When the wall of the distended artery fails to support the stress occurs rupture of the aneurysm. This rupture often leads to death or severe disability.

This chapter introduces analysis about pathogenesis, hemodynamic forces acting on the wall vessel that could be important factor to the origin and progression of the disease, and computational fluid dynamic (CFD) associated to the medical imaging. Thus, the interaction between pathogenesis of aneurysm, medical imaging and CFD are important to understand the development of aneurysm.

2. Pathogenesis of arterial aneurysms

2.1 General considerations

An arterial aneurysm is one of the most common vascular diseases causing disability and death. True aneurysm represents a degeneration of the artery wall with loss of structural integrity leading to gradual dilatation of all artery wall layers. Aneurysms have been reported in almost all segments of the arterial tree, but are more frequently localized at the aorta below the renal arteries. Aneurysms are most commonly diagnosed in the sixth and seventh decades of life, with a rising incidence for unknown reasons. The prevalence of aneurysms in a given population depends on the presence of risk factors associated, including older age, male gender, white race, positive family history, smoking, hypertension, hypercholesterolemia, peripheral vascular occlusive disease, and coronary artery disease. Although these risk factors are associated with increased abdominal aortic aneurysms (AAA), they may not be independent predictors and may be markers rather than causes of AAA prevalence.

Although a precise cause of AAA remains unknown, much has been learned about the pathophysiology of the aneurysmal aorta. Research has linked the development of AAA with chronic aortic wall inflammation, increased local expression of endogenous proteinases, and the degradation of structural connective tissue proteins (Shah, 1997).

Most arterial aneurysms arise at the bifurcation of major arteries, and this is also true for the intracranial location. Around 85% of all intracranial aneurysms originate from the anterior circulation. The most common location (30%–35%) is the anterior communicating artery. The prevalence of intracranial aneurysms among first-degree relatives of patients with cerebral aneurysms is higher than in the general population. The risk for a first-degree relative harbouring an aneurysm is about three to four times higher than for someone from the general population (Raaymakers 1999; Ronkainen et al. 1997).

Although the pathogenesis and etiology of cerebral aneurysms has been studied extensively, both are still poorly understood. Endogenous factors like elevated blood pressure, the special anatomy of the Circle of Willis or the effect of hemodynamic factors, particularly originating at vessel bifurcations, are all known to be involved in the growth and rupture of an aneurysm. Arteriosclerosis and inflammatory reactions, however, might also have an impact. Exogenous factors like cigarette smoking, heavy alcohol consumption or certain medications are thought to be risk factors in the pathogenesis of an aneurysm or at least increase the risk of rupture.

2.2 Basic mechanisms

The normal aortic wall is composed of a thin endothelial lined intima, a thick elastic media dominated by vascular sooth muscle cells, and a fibrocollagenous adventitia. Medial elastin fibers and interstitial collagens are normally responsible for the tensile strength, resilience, and structural integrity of the aortic wall. Aneurysmal dilatation and rupture are due to mechanical failure of these fibrillar extracellular matrix proteins.

Basic research on the pathophysiologic processes of aortic aneurysm formation and growth has identified several putative mediators of the disease. These mediators include both bacteriologic and enzymatic agents, especially matrix metalloproteinases. Aneurysm formation and aneurysm growth represent two distinct phases of aortic aneurysm disease. Aneurysm formation encompasses the metabolic processes that cause degradation of the structural components of the aortic wall and loss of biomechanical integrity. This phase of aneurysm disease appears to be directed at the destruction of elastin within the aortic wall. Although all layers of the aortic wall may be involved, the most fundamental structural alteration that results in a loss of biomechanical integrity and aneurysm formation is degradation of adventitial elastin. The metabolic processes that cause adventitial elastin degradation are nonselective. They most likely begin in the intima or media, associated with an inflammatory cell infiltrate. Destruction of the elastin within these layers does not cause loss of the structural integrity of the aortic wall. These processes, however, ultimately invade the adventitia, cause degradation of the elastin within this layer, destroy the structural integrity of the aortic wall, and permit its pathologic expansion. Because humans are unable to synthesize and deposit elastin in the aortic wall in any detectable quantity beyond the first year of life, the degraded elastin is replaced by collagen I and III during the process of aneurysm formation. (Mesh et al., 1992) Once formed, even the smallest aneurysm represents the end-stage of aortic wall destruction, a significant alteration in the biomechanical properties of the aortic wall and a reduction in structural integrity. Aneurysm growth appears to be a self-sustaining process. Data from both human and animal aneurysm cells demonstrate a sustained increase in collagen types I and III m RNA and new collagen deposition within the aneurysm wall. These processes continue until the rate of collagen degradation exceeds the rate of deposition or the pressure per unit area of aneurysm wall exceeds the ability of the collagen fibers to withstand the load. In either of these situations, aneurysm rupture occurs.

3. Evaluation of aneurysms

3.1 Physical examination and imaging

Although most clinically significant AAA are potentially palpable during routine physical examination, the sensitivity of this technique depends on size, obesity of the patient, skill of the examiner, and focus of the examination. With physical examination alone, the diagnosis is made in 29% of AAAs 3 to 3.9 cm, 50% AAAs 4 to 4.9 cm, and 75% of AAAs 5 cm or larger (Lederle & Simel, 1999).

Several imaging modalities are available to confirm the diagnosis of AAA. Abdominal B-mode ultrasonography is the least expensive, least invasive, and most frequently used examination, particularly for initial confirmation of suspected AAA and follow up of small AAAs. Computed tomography is more expensive than ultrasound and involves exposure to radiation and intravenous contrast material, but it provides more accurate measurement of diameter, with 91% of studies showing less than 5 mm interobserver variability (Chervu et al., 1995).

3.2 Factors affecting clinical decision making

The choice between observation and prophylactic surgical repair of an AAA for an individual patient at any given time should take into account the risk for rupture under observation, the operative risk associated with repair, the patient's life expectancy, and the personal preferences of the patient. For the scope of this chapter we will only analyse the risk of rupture.

Estimates of the risk for AAA rupture are imprecise because large numbers of patients with AAAs have not been observed without intervention. Data are insufficient to develop an accurate prediction rule for AAA rupture in individual patients, which makes surgical decision making difficult. Some aspects may be considered:

- Aneurysm diameter

For the past 5 decades maximal aneurysm diameter has been the primary determinant of rupture risk. Several studies firmly established the effect of size on AAA rupture and provided a sound basis for recommending elective repair for large AAAs because of marked improvement in survival after repair. Despite differences in precise estimates, most studies show that rupture risk increase substantially with AAA diameter between 5 – 6 cm.

- Aneurysm wall stress

From a biomechanical perspective, AAA rupture occurs when the forces within a AAA exceed the wall's "bursting strength. The application of engineering principles to the analysis of actual aneurysms has only recently been possible. Multiple studies have demonstrated that finite element analysis of AAA wall stress with three dimensional CT reconstructions is better than diameter for estimating rupture risk.

- Finite element analysis

At this point the strength of the data and the size of the patient cohorts already rival or exceed that of the data initially used to determine the clinical use of aneurysm diameter to estimate rupture risk in the 1960's. The technique of using aneurysmal wall stress to predict rupture risk remains to be validated in a large multicenter cohort using a standardized, broadly applicable technique, although one such study is currently under way (Fillinger et al., 2004).

- Aneurysm shape

Clinical opinion holds that shape is important and eccentric or saccular aneurysms present a greater risk for rupture than do more diffuse fusiform aneurysms. Vorp et al. (1998) associates using computer modelling showed that wall stress is substantially increased by an asymmetric bulge in AAA. The presence of calcification in the wall may increase wall stress focally, but may not be useful as a clinical tool. The effect of intraluminal thrombus on rupture risk is also debated, with studies suggesting that thicker thrombus may increase the risk for rupture, decrease the risk for rupture, or have no effect. The practical impact of these variables on AAA rupture risk requires further study.

4. Biomechanical analysis

This section provide an overview of the fundamental basis of biomechanical factors acting on the vessel wall and their influences under endothelial cells and the development of the aneurysm, once that there are several studies showing disturbed flow conditions and unsteady turbulent stresses damage endothelial cells and may provide a first step to the degradation of the wall (Davies et al., 1984, 1995). There will be analysed biomechanical factors in intracranial aneurysm and an abdominal aortic aneurysm.

4.1 Intracranial aneurysm

At present, there is not completely satisfied the theory about the origin, growth and rupture of intracranial aneurysm. The walls of intracranial arteries exhibit the same general organization and composition of all arteries. However, in most cerebral arteries, there is no external elastic lamina. Moreover, in these arteries, the arterial walls are thin, there is less elastin, and there are "medial defects" (gaps in the muscle layer) present frequently, thus intracranial arteries are more susceptible to aneurysm formation than extracranial arteries. Other arteries show similar medial defects, particularly in renal, mesenteric, splenic, and coronary arteries. However, these defects are smaller and less frequent and the aneurysm formation is rarely (Crompton, 1966). In the literature, there are some hypothesis about the location and structural changes in the wall of the artery that contribute to development of the aneurysm. Eppinger (1887)(apud Thubrikar, 2007) showed that certain aneurysms occurred at the site of the medial defects, however, other researchers, such as, Forbus (1930), Toth et al. (1998) and Zhang et al. (2003) showed that the aneurysm could be developed through combination of degeneration of the elastic and defects in the muscular layer. Meng et al. (2007) recently demonstrated in animal models that the aneurysm initiation and development occurred at the apices of arterial bifurcations where there was high wall shear stress and aneurysm-type wall remodelling (disrupted internal elastic lamina and endothelium, thinned media and smooth muscle cells loss) at histology. These factors appear to render these arteries susceptible to a local weakening under the persistent action of hemodynamic loads, particularly in hypertension (Inci and Spetzler, 2000). The hemodynamic loads resulting from the direct impingement of the central streams at the apex of bifurcations are probably the most important factor contributing to the focal degeneration of the internal elastic membrane and the early origin the aneurysms. These forces could enlarge medial defects already present. The impingement of central axial streams results in a much greater velocity gradient and shear stress at the apex than is experienced in the main stem or branches of bifurcations. As flow is pulsatile, the peak force will be great, because there is a brief impact time and in the moment of impact the kinetic energy of the moving blood is changed to pressure energy (stagnation pressure) (Thubrikar, 2007). This extra pressure is the force responsible for focal degeneration of the internal elastic membrane and thus the cause of initiation of aneurysms at the apex. Shojima et al. (2005) reported that local rises in pressure due to flow impingement are less than 2 mmHg, which is small compared with nominal pressure levels in cerebral arteries, and concluded that dynamic pressures acting at bifurcations and on the walls of intracranial arteries may be less significant to enlargement and rupture than previously assumed. Acevedo-Bolton et al (2006) concluded that regions that continued to enlarge experienced low wall shear stress and speculated that this might be due to increased residence time of particles that degrade the aneurysm wall. There is a great interest to study the wall shear stress in aneurysms because there are many works show that it plays a role in the evolution of aneurysmal disease. It is known that the wall shear stress is a major physiological stimulus for the vessel endothelium. The endothelial cells submitted the physiological levels of wall shear stress in the arteries [1 (10) – 7 (70) Pa (dyn/cm^2)] and in the veins (0.1 (1) – 0.7 (7) Pa (dyn/cm^2)] provide a physiological stimulus for the vessel endothelium contributing to selective barrier for macromolecular permeability, can influence vascular remodelling via the production of growth-promoting and –inhibiting substances, modulate hemostasis/thrombosis through the secretions of procoagulant, anticoagulant, and fibrinolytic agents, mediate inflammatory responses via the surface expression of chemotactic and adhesion molecules and release of chemokines and cytokines (Malek et al., 1999; Paszkowiak & Dardik, 2003; Li et al., 2005). On the other hand, when there is excess or lack of the stimulus can lead to pathological phenomena that cause changes in the arterial wall biomechanical properties. The high values of wall shear stress can cause damage in the endothelial cells (Fry, 1968). A prolonged high wall shear stress fragments the internal elastic lamina of vessels (Masuda et al., 1999) and gives rise to the initial change involved in the formation of cerebral aneurysm. Low wall shear stress has been reported to be related to aneurysmal growth (Jou et al., 2003) and rupture (Shojima et al., 2004) as it promotes various mechanisms that cause arterial wall remodelling. When the wall shear stress is lower than 0.4 Pa, it generates endothelial proliferation (Malek et al., 1999) and apoptosis (Kaiser et al., 1997). It is also responsible for abnormal vascular reactivity and vasospasm that can cause ischemia, angina, and myocardial infarction, increased permeability to macromolecules such as lipoproteins, increased expression of chemotatic molecules and adhesion molecules, enhanced recruitment and accumulation of monocytes/macrophages in the intima as foam cells, altered regulation in growth and survival of vascular cells (Lerman & Burnett, 1992; Bonetti et al., 2003; Gimbrone et al., 2000). This excessive low wall shear stress may be one of the main factors underlying the degeneration, indicating the structural fragility of the aneurysmal wall (Shojima et al., 2004). It appears that, after an initial injury that might result from excessive wall shear stress on the endothelial cells (Meng et al., 2007), progressive changes in aneurysm shape occur with a trend for the cross section to become more elliptical (Boussel et al., 2008). This generates a progressive decrease of wall shear stress leading to endothelial dysfunction, wall remodelling, and aneurysm growth (Ahn et al., 2007; Utter & Rossman, 2007).

4.2 Abdominal aortic aneurysm

Abdominal aortic aneurysms (AAAs) most of the affect men between 40 and 70 years old (5-7% of people older than 60) (Thiriet, 2008). AAAs rarely appear in individuals under 50 years old, but their incidence increases drastically at age 55 and peaks in the early 80s. Norway in 1994-1995 showed that AAAs are present in 8.9% of men and in 2.2% of women over 60 years old (Singh et al., 2001). Reed et al. (1992) showed that AAAs increased steadily with age after 60 years old and aortic dissections (aortic wall dissects and blood enters the wall causing the enlargement) showed peak between ages 70 and 75 years old and decreased after that (Figure 2).

Although aneurysms may develop throughout ht length of the aorta,   AAAs are at least 5 times more prevalent than thoracic or thoracoabdominal aneurysms (Dua & Dalman, 2007). This fact can be explained from two points: physiologic and anatomic features unique to the distal aorta. The infrarenal aorta is the most common site of extracranial aortic aneurysm formation. Hemodynamic through the aorta combined with regional factors can explain this preferential distribution. In the infrarenal aorta the number of elastic lamellae (and therefore elastin) is markedly decreased in comparison with the thoracic aorta which become fragmented and unorganized (Lakatta et al., 1987; MacSweeney et al., 1994) contributing to reducing elasticity and wall motion (Ailawadi et al., 2003). The degeneration of elastic fibers is accompanied by an increase in the collagenous substance (the stiffer structural component). As the ratio of elastin to collagen decreases, the vessel progressively loses its elasticity. The stiffening of the wall causes an increase in speed of the pulse wave. For example, in the aorta, the wave speed increases from 6.5 m/s in a 10-year-old child to upwards of 11 m/s in a 60-year-old adult (Nichols & O'Rourke, 1990). Reduced distal aortic elasticity, in combination with augmented pressure due to pulse wave reflections from the aortic bifurcation and other downstream arteries, may increase wall strain and aneurysm susceptibility (Humphrey and Taylor, 2008). The collagen-to-elastin ratio is the principal determinant of wall mechanics in the aorta. Changes in composition and structure of the arterial wall will alter the wall mechanics. An increase in collagen-to-elastin ratio results in a higher wall stiffness and lower tensile strength. Clinical observations show that most AAA walls become progressively stiffer as the diameter increases. This is because of biomechanical restructuring of the wall (Kleinstreuer et al., 2007). In the normal abdominal aorta, the collagen-to-elastin ratio is approximately 1.58 (Nichols and O'Rourke, 1990), however, the collagen-to-elastin ratio is much higer in AAAs (table 1).

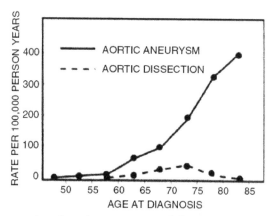

Fig. 2. Plots of incidence rates of aortic aneurysm and dissection by age at diagnosis. Extracted from Reed et al., (1992).

As commented above, hemodynamic of the infrarenal is one the main factors to cause the development of AAAs. Hemodynamic forces relevant to AAA pathogenesis can be obtained into three components: 1) wall shear stress (explained above), 2) hydrostatic pressure, the perpendicular force acting on the vascular wall; and 3) relative wall strain (RWS), the circumferential stretch of the vessel wall exerted by cyclic luminal pressure changes and resulting tensile stress. Several works in the literature showed that cultured vascular endothelial cells studies when submitted the disturbed flow conditions and unsteady turbulent stresses damage the endothelium, and the loss or malfunctioning of their regulatory processes may provide a first step to the degradation of the wall (Davies et al. 1984, 1995, 2009; Chiu and Chien, 2011). Hemodynamic conditions vary markedly along the aorta, from high Reynolds numbers (Re)

$$Re = \frac{\rho V d}{\mu} \tag{1}$$

where ρ is the specific mass, V is media velocity, d is the diameter of the vessel and μ is the viscosity dynamic, at the aortic root to low and oscillatory shear conditions at the aortic bifurcation (Greve et al., 2006). Most relevant to AAA disease pathophysiology, and its predilection for the distal-most aortic segment, is the marked difference between resting aortic wall shear stress in the thoracic and abdominal aorta. In suprarenal aortic segments, flow is antegrade throughout the cardiac cycle, providing continuous antegrade laminar wall shear stress. In infrarenal aorta, wall shear stress values are lower, and reverse flow is present in late systole and diastole. In response to reduced distal arterial resistance and increase flow, such as is demonstrated in the response to even modest lower extremity exercise, wall shear stress becomes antegrade and laminar throughout the cardiac cycle, mimicking those characteristic of more proximal aortic segments. These distinct regional differences in hemodynamic influences may account for some component of the differential aneurysm risk noted between the thoracic and abdominal aortic segments (Dua and Dalman, 2010).

	Normal Aorta	Aneurysm
Elastin		
Average	22.7	2.4
Maximum	32.5	6.7
Minimum	16.1	0.2
Muscle		
Average	22.6	2.2
Maximum	33.6	6.4
Minimum	15.5	0.4
Collagen and ground substances		
Average	54.8	95.5
Maximum	63	98
Minimum	48	91.4

Table 1. Composition of normal aorta and aneurysm. Extracted from Nichols & O'Rourke, 1990.

As commented above, there are several work in the literature that show a correlation between very low shear stresses and the loss of permeability of the endothelial cell membrane (Helmlinger et al., 1991; Chiu et al., 2003). Studies have been the basis of an alternative, or even complementary, mechanism responsible for the origin of these aneurysms. During the normal course of aging, the abdominal aortic artery gradually undergoes conformal changes in its geometry (increasing its length and diameter, thickening its wall, etc.). Over time, the relative unconstructed nature of this artery inside the abdominal cavity may lead to the formation of bends, kinks, and other morphological changes that, in turn, create "disturbed flow" conditions inside the vessel (i.e., unsteady flow separation and weak turbulence). It is the argued that the anomalous response of the VEC to the high shear stresses, very low shear stresses, low, but oscillating shear stresses, and the anomalous temporal and spatial gradients of wall shear stress associated with these disturbed flow conditions could contribute to an unstable progressive degradation of the arterial wall and to the formation of the aneurysm (Lasheras, 2007). To understand more about the risk for and progression of aneurysm disease is necessary introduce DFC to analyse hemodynamic in site specific, for example, in the aorta infrarenal. Compared with the suprarenal aorta, the infrarenal environment in resting subjects is characterized by increased peripheral resistance, increased oscillatory wall shear stress and stagnant flow (Dua and Dalman, 2010).

5. Computational fluid dynamics (CFD)

Since the beginning of the computer age, the computational study of fluid dynamic problems has been of interest to researchers studying both fundamental problems and engineering applications. Vast numbers of real-world problems require accurate viscous flow solutions to meet requirements for supporting engineering and science tasks – such as achieving ideal fluid dynamic performances and satisfying cost effectiveness. For example, computational analysis is indispensable, as well as economical, for developing advanced rocket-engine turbopumps and biomedical devices handling blood flow in humans. The computational fluid dynamics (CFD) for viscous, incompressible flow has been of interest for many decades to investigate fundamental fluid dynamic problems as well as engineering applications. The pioneering work by Harlow and Welch (1965) opened a new possibility of applying a computational approach to solving realistic incompressible fluid engineering problems, especially for three-dimensional problems (Kwak and Kiris, 2011).

Development of image-based modelling technologies for simulating blood flow began in the late 1990s. Since that time, many groups have developed and utilized these techniques to investigate the pathogenesis of occlusive and aneurysmal disease in the carotid artery (Long et al., 2000), the coronary arteries (Gijsen et al., 2007), the aorta (Tang et al., 2006) and the cerebral circulation (Cebral et al., 2005). Patient-specific modelling techniques have also been applied in solid mechanics analyses to predict rupture risk of aneurysms (Vorp, 2007).

There are many methods for quantifying vascular anatomy for patient-specific modelling of cardiovascular mechanics include noninvasive imaging techniques such as computed tomography (CT), magnetic resonance imaging (MRI), 3D ultrasound (3DUS) and an invasive method combining angiography and intravascular ultrasound (IVUS). Contrast-enhanced CT and MRI are particularly suited for generating high-resolution volumetric images of many parts of the vascular tree. Generally, iodinated contrast is used in CT angiography, and a gadolinium-based contrast agent is used in magnetic resonance (MR) angiography. MRI has the additional advantage of being able to quantify physiologic parameters, including blood flow, wall motion and blood oxygenation (Taylor & Figueroa, 2009).

A recent serial MRI-based case study demonstrated an association between the eventual site of plaque ulceration and elevated wall shear stress (WSS) (Groen et al., 2007) , a finding corroborated by an IVUS-based CFD study of 31 plaques that showed a clear association between elevated WSS and elevated strains within the plaque (Gijsen et al., 2008). Tang et al. (2006) quantified hemodynamic conditions in subject-specific models of the human abdominal aorta constructed from magnetic resonance angiograms (MRA) of five young, healthy subjects. Image-based modelling techniques are also being used to provide hemodynamic data in a clinical study testing the hypothesis that exercise can be employed to slow the progression of small (3-5 cm diameter) abdominal aortic aneurysms (Dalman et al., 2006). Rayz et al. (2008) performed a study on the effect of vertebral artery flow in four patients with basilar artery aneurysms. Computed flow fields were found to agree well with measurements made using PC-MRI. Subsequently, these authors reported good correspondence between regions of slow flow predicted by CFD at baseline and the deposition of thrombus observed at follow-up of three basilar aneurysms cases.

5.1 Real model of artery

One way to analyze and estimate the hemodynamic behaviour in an artery is the use of computational analysis of a real model of blood vessel (Legendre, 2010). The artery model is built based on images obtained from a patient. For example, in order to evaluate an abdominal aortic aneurysm in a patient, this model can be obtained by using multi-slice CT scan of the patient chest (Figure 3).

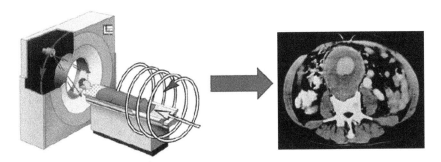

Fig. 3. CT images acquisition of a patient. Extracted from Invesalius (2008).

These DICOM (Digital Image Communication in Medicine) images should be treated by dedicated software, for example InVesalius, Mimics, etc., to select only the region of interest. Then, it is generated a compatible extension file that can be imported by a software to generate a three-dimensional mesh model of the arterial segment under study.

Overlapping the two dimensions (2D) images obtained through computed tomography or MRI, it can be created the computational model in three dimensions (3D) corresponding to the patient anatomical structure (Figure 4) and (Figure 5).

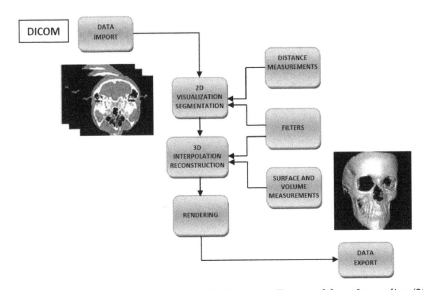

Fig. 4. Flowchart of the reconstruction of medical images. Extracted from Invesalius (2008).

The program makes, from tomographic medical imaging, a three-dimensional computational model in genuine size of several anatomical structures. This model can be manipulated and observed from different angles and it is also possible to separate specific parts of the image obtained from CT for a detailed analysis. The different densities of bone and tissue are easily identified by using tools such as color map. This feature is an important tool for creating computational models. Different methods of volume segmentation can be used for the treatment of the obtained images. Those methods are useful to adequately separate the regions of tissues, organs, anatomical structures etc.

The computational model generated by the software enables computer simulations in order to evaluate hemodynamic patterns developed by each patient according to their individual characteristics (Figure 6).

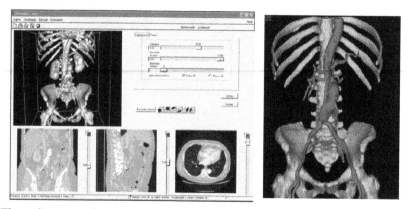

Fig. 5. Three-dimensional reconstruction of the abdominal aorta segment obtained from CT Scan. Extracted from Invesalius (2008).

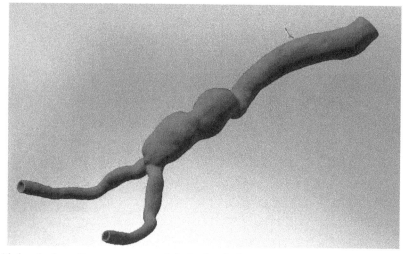

Fig. 6. Abdominal aortic aneurysm model obtained after image processing. Extracted from Legendre (2010).

5.2 Computer analysis
5.2.1 Finite volume method
The finite volume method (FVM) is a numerical technique for the solution of partial differential equations. The domain under study is subdivided into control volumes or computational cells. This method consists in the integration of the equations that governing the fluid flow over each control volume. The numerical solutions obtained by techniques of FVM (Patankar, 1975) have problems or errors known as false numerical diffusion. The numerical diffusion occurs when the interpolation function used in the discretization of equations differs from the exact solution. It can be understood as any effect that tends to moderate gradients or discontinuities present in the exact solution of a problem. There are some functions that were developed in order to minimize these effects of false diffusion, for example, the "upwind" method of second order.

For instance, the Fluent software solve the governing integral equations for the conservation of mass and momentum and also for scalars such as turbulence. The technique based on the volume control works as the following:
- The domain is divided in discrete control volumes using a computational mesh;
- Integration of the governing equations of the individual control volume to construct algebraic equations for discrete dependent variables (unknown), such as speed, pressure, etc.;
- Linearization of the discretized equations and solution of the resulting linear equation system to update the field values of the dependent variables.

5.2.2 Segregation solution method
In the segregated solution method, the governing equations are solved sequentially. Thus, to solve the equations that governing the phenomenon, many interactions are carried out until the solution convergence. Each interaction consists of the following steps outlined below:
- The fluid properties are updated based on the current solution. If the calculation has been started, the fluid properties are updated based on the initialized solution;
- The equations of momentum in x, y and z are solved one by one using the current values of pressure and mass flow to update the velocity field;
- Since the velocities obtained in step 2 cannot satisfy the conditions of continuity, pressure corrections are made so that continuity is satisfied;
- Other equations for scalar quantities such as turbulence are solved using the previously updated value of the other variables;
- It is made a convergence confirmation of the equations set.

These steps are performed until the convergence criteria are achieved.

5.2.3 Discretization
The technique based on control volume converts the governing equations into algebraic equations so that they can be solved numerically. This control volume technique consists of integrating the governing equations about each control volume, generating discrete equations that conserve each quantity on the volume control.

5.3 Overview
In general, the computational analysis tool can be viewed as a data interpolation in which the uncertainties of the results will depend on the model, the boundary conditions and mesh refinement adopted. To perform a study for hemodynamic performance evaluation, the

computational model adopted from the arterial segment should be able to represent the real model in a reliable way, as much as possible. The boundary conditions should be properly employed, since they are directly related to the consistency of the numerical results. The use of the finite volume method for solving the problem requires a proper discretization of the model in order to enable the convergence of analysis and also minimize the uncertainties in the results. On the other hand, it is also needed a good sense to perform the mesh refinement and computational cost in order to obtain a good quality of results in the shortest time. The quality of the mesh has an important role in the accuracy and stability of numerical computation. The convergence of results is intrinsically related to the size of mesh elements of the model and thus its quality is measured by its "asymmetry".

The choice of the optimal mesh for the simulation of the model should consider the performance and accuracy. The performance will depend on the total number of elements to be analyzed and thus a computational mesh made by larger elements will require lower computational cost due to lower number of elements required to cover the whole region of interest. However, this may affect the consistency of results. On the other hand, smaller mesh elements means better results, but this will require a greater computational cost (Figure 7).

One way to obtain more information regarding the arterial hemodynamic behavior is through the use of computer simulation. By using this kind of tool, it is possible to evaluate wall stress, recirculation areas, pressure, pathlines, velocity field and speeds distribution in the arterial segment.

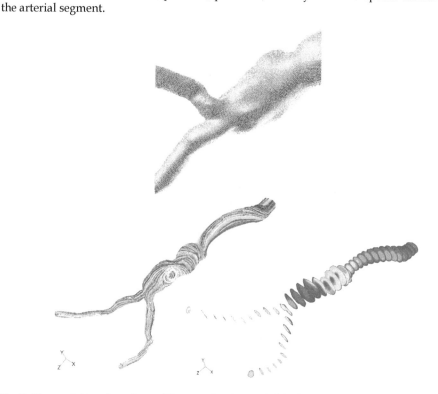

Fig. 7. Computational mesh, pathlines and velocity field of an arterial aneurysm. Extracted from Legendre (2010).

There are many works numerical and experimental fluid mechanics in the literature aimed at determining the characteristics of the flow shear stresses on the walls of AAAs at different stages of their development. These studies occur in ideal symmetric and nonsymmetric shapes of fusiform aneurysms and in realistic geometries obtained from high-resolution CT and angiographies (Egelhoff et al. 1999, Finol & Amon 2003, Salsac et al. 2006, Bessa & Ortiz 2009, Legendre 2010). Although, these studies show limitations about the appropriate initial and boundary conditions, as well as account for the precise elastic properties of the wall, they clearly show that once a fusiform aneurysm forms, the flow is dominated by the onset of an unsteady, massive separation from the walls that occurs immediately after the peak systole. When the flow separates from the walls during the deceleration portion of the cardiac cycle, a relatively coherent array of large vortices forms and the blood flow slowly recirculates (Lasheras, 2007). Bessa & Ortiz (2009) showed that occur massive separation from the walls and the velocity profile becomes inverted inside of the aneurysm, forming a vortex in the bottom and in the top of the aneurysm. After that, the velocity profile has been completely inverted and the vortex pair travels towards to the center of the aneurysm (Figure 8).

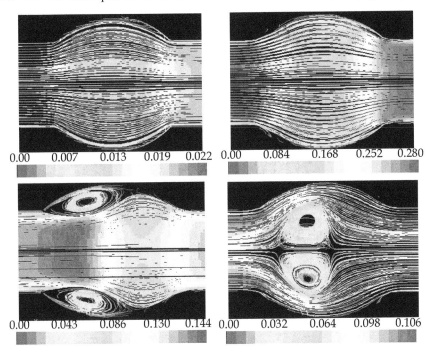

Fig. 8. Streamlines at various instant time of the cardiac cycle for aneurysm model. Extracted from Bessa & Ortiz (2009).

6. Conclusion

In the last few decades, it had a great effort of the scientific community to identify the beginning and the progression of the aneurysms. For this, several visualization techniques and 3D reconstruction of the vessels were developed associates with the computational fluid dynamics with intention to analyze hemodynamic parameters that contributed to the development of the aneurysms. However, many questions exist on aneurysms that continue without answers: What cause aneurysms? When occur will the rupture of the aneurysms? Nowadays, advances in computational methods and computing hardware are also making it possible to solve increasingly more challenging problems, notably those involving flow instabilities and turbulence associated with a variety of vascular pathologies. However, to advance the research is the necessary to improve the related uncertainties the inserted boundary conditions in computational model. Exactely, even if these issues are fully resolved, still the mechanotransduction knowledge is essential to understand the endothelial cells behaviour. Therefore, still it has great challenges for the understanding of the origin and development of the aneurysm.

7. Acknowledgment

This chapter was supported, in part, by grants from the Conselho Nacional de Desenvolvimento Científico e Tecnológico (CNPq) and Fundação de Amparo a Pesquisa do Estado de São Paulo (FAPESP).

8. References

Acevedo-Bolton, G., Jou, L.D., Dispensa, B.P., Lawton, M.T., Higashida, R.T., Martin, A.J., Young, W.L. & Saloner, D. (2006). Estimating the Hemodynamic Impact of Interventional Treatments of Aneurysms: Numerical Simulation with Experimental Validation: Technical Case Report. *Neurosurgery*, Vol. 59, pp. E429-E430.

Ahn, S., Shin, D., Tateshima, S., Tanishita, K., Vinuela, F. & Sinha, S. (2007). Fluid-Induced Wall Shear Stress in Anthropomorphic Brain Aneurysm Models: Mr Phase-Contrast Study at 3 T. *Journal of Magnetic Resonance Imaging*, Vol. 25, pp. 1120-1130.

Ailawadi, G., Eliason, J. & Upchurch Jr., G. (2003). Current Concepts in the Pathogenesis of Abdominal Aortic Aneurysm. *Journal of Vascular Surgery*, Vol. 38, 584-588.

Bessa, K.L. & Ortiz, J.P. (2009). Flow Visualization in Arteriovenous Fistula and Aneurysm Using Computational Fluid Dynamics. *Journal of Visualization*, Vol. 12, pp. 95-107.

Bessa, K.L., Belletati, J.F., dos Santos, L., Rossoni, L.V. & Ortiz, J.P. (2011). Drag Reduction by Polyethylene Glycol in the Tail Arterial Bed of Normotensive and Hypertensive Rats. *Brazilian Journal of Medical and Biological Research*, Vol. 44, pp. 767-777.

Bonetti, P.O., Lerman, L.O. & Lerman, A. (2003). Endothelial Dysfunction: A Marker of Atherosclerotic Risk. *Arteriosclerosis, Thrombosis and Vascular Biology*, Vol. 23, pp. 168-175.

Boussel, L, Rays, V., McCulloch, C., Martin, A., Acevedo-Bolton, G., Lawton, M., Higashida, R., Smith, W.S., Young, W.L. & Saloner, D. (2008). Aneurysm Growth Occurs at Region of Low Wall Shear Stress: Patient-Specific Correlation of Hemodynamics and Growth in a Longitudinal Study. *Stroke*, Vol. 39, pp. 2997-3002.

Cebral, J.R., Castro, M.A., Burgess, J.E., Pergolizzi, R.S., Sheridan, M.J. & Putman, C.M. (2005). Characterization of Cerebral Aneurysms for Assessing Risk of Rupture by Using Patient-Specific Computational Hemodynamics Models. *American Journal of Neuroradiology*, Vol. 26, pp. 2550-2559.

Chervu, A., Clagett, G.P., Valentine, R.J., Myers, S.I. & Rossi, P.J. (1995). Role of Physical Examination in Detection of Abdominal Aortic Aneurysms. *Surgery*, Vol. 117, pp. 454-457.

Chiu, J.J. & Chien, S. (2011). Effects of Disturbed Flow on Vascular Endothelium: Pathophysiological Basis and Clinical Perspectives. *Physiological Reviews*, Vol. 91, pp. 327-387.

Chiu, J.J., Chen, L.J., Lee, P.L., Lee, C.I., Lo, L.W., et al. (2003). Shear Stress Inhibits Adhesion Molecule Expression in Vascular Endothelial Cells Induce by Coculture with Smooth Muscle Cells. *Blood*, Vol. 1, pp. 2667-2674.

Crompton, M.R. (1966). The Pathogenesis of Cerebral Aneurysms. *Brain: a journal of neurology*, Vol. 4, (December), pp. 797-814.

Dalman, R.L., Tedesco, M.M., Myers, J. & Taylor, C.A. (2006). AAA Disease: Mechanisms, Stratification, and Treatment. *Annals of the New York Academy of Sciences*, Vol. 1085, pp. 92-109.

Davies, P.F. (2009). Hemodynamic Shear Stress and the Endothelium in Cardiovascular Pathophysiology. Natural Clinical Practice. *Cardiovascular Medicine*, Vol. 6, pp.16-26.

Davies, P.F., Dewey, C.F., Bussolari, S., Gordon, E. & Gibrone, M.A. (1984). Influence of Hemodynamics Forces on Vascular Endothelial Function. *The Journal of Clinical Investigation*, Vol. 4 (April), pp. 1121-1129.

Davies, P.F., Mundel, T. & Barbee, K.A. (1995). A Mechanism for Heterogeneous Endothelial Responses to Flow in vivo and in vitro. *Journal of Biomechanics*, Vol. 28, (December), pp. 1553–1560.

Dua, M.M. & Dalman, R.L. (2010). Hemodynamic Influences on Abdominal Aortic Aneurysm Disease: Application of Biomechanics to Aneurysm Pathophysiology. *Vascular Pharmacology*, Vol. 53, pp. 11-21.

Egelhoff, C.J., Budwig, R.S., Elringer, D.F., Khraishi, T.A. & Johansen, K.H. (1999). Model Studies of the Flow in Abdominal Aortic Aneurysms During Resting and Exercise Conditions. *Journal of Biomechanics*, Vol. 32, pp. 1319-1329.

Eppinger, H. (1887). Pathogenesis (Histogenesis und Aetiologie) der Aneurysmen einschliesslich des Aneurysma equi verminosum. Pathologisch-anatomische studien. Arch Klin Chir, Vol 35, pp. 1-563.

Fillinger, M.F., Racusin, J., Baker, R.K., Cronenwett, J.L., Teutelink, A., Schermerhorn, M.L., Zwolar, R.M., Powell, R.J., Walsh, D.B., Rzucidlo, E.M. (2004). Anatomic Characteristics of Ruptured Abdominal Aortic Aneurysm on Conventional CT Scans: Implications for Rupture Risk. *Journal of Vascular Surgery*, Vol. 39, pp. 1243-1252.

Finol, E.A. & Amon, C.H. (2003). Flow Dynamics in Anatomical Models of Abdominal Aortic Aneurysms: Computational Analysis of Pulsatile Flow. *Acta Científica Venezolana*, Vol. 54, pp. 43-49.

Forbus, W.D. (1930). On the Origin of Miliary Aneurysm of the Superficial Cerebral Arteries. Bull Johns Hopkins Hospital, vol 47, pp 239-284.

Fry, D.L. (1968). Acute Vascular Endothelial Changes Associated with Increased Blood Velocity Gradients. *Circulation Research*, Vol. 22, pp. 165-197.

Gijsen, F.J., Wentzel, J.J., Thury, A., Lamers, B. & Schuurbiers, J.C. (2007). A New Imaging Technique to Study 3-D Plaque and Shear Stress Distribution in Human Coronary Artery Bifurcations in ViVo. *Journal of Biomechanics*, Vol. 40, pp. 2349-2357.

Gijsen, F.J., Wentzel, J.J., Thury, A., Mastik, F., Shaar, J.A., Schuurbiers, J.C., Slager, C.J., van der Giessen, W.J. & de Feyter, P.J. (2008). Strain Distribution Over Plaques in

Human Coronary Arteries Relates to Shear Stress. *American Journal of Physiology. Heart and Circulatory Physiology*, Vol. 295, pp. H1608-H1614.

Gimbrone, M.A. Jr., Topper, J.N., Nagel, T., Anderson, K.R. & Garcia-Cardena, G. (2000). Endothelial Dysfunction, Hemodynamic Forces, and Atherogenesis. *Annals of the New York Academy of Sciences*, Vol. 902, pp. 230-240.

Greve, J.M., Les, A.S., Tang, B.T., Draney Blomme, M.T., Wilson, N.M., Dalman, R.L., Pelc, N.J. & Taylor, C.A. (2006). Allometric Scaling of Wall Shear Stress from Mice to Humans: Quantification using Cine Phase-Constrast MRI and Computational Fluid Dynamics. *American Journal of Physiology. Heart and Circulatory Physiology*, Vol. 291, pp. H1700-H1708.

Groen, H. C., Gijsen, F.J., van der Lugt, A., Ferguson, M.S., Hatsukami, T.S., van der Steen, A.F., Yuan, C. & Wentzel, J.J. (2007). Plaque Rupture in the Carotid Artery is Localized at the High Shear Stress Region: a Case Report. *Stroke*, Vol. 38, pp. 2379-2381.

Harlow, F.H. & Welch, J.E. (1965). Numerical Calculation of Time-Dependent Viscous Incompressible Flow with Free Surface. *The Physics of Fluids*, Vol. 8, pp. 2182-2189.

Helmlinger, G., Geiger, R.V., Schrech, S. & Nerem, R.M. (1991). Effects of Pulsatile Flow on Cultured Vascular Endothelial Cell Morphology. *Journal of Biomechanical England*, Vol. 113, pp. 123-131.

Hisai, T.K. (2008). Mechanosignal Transduction Coupling between Endothelial and Smooth Muscle Cells: Role of Hemodynamics Forces. *American Journal of Physiology Cell Physiology*, 294, C659-661.

Humphrey, J.D. & Taylor, C.A. (2008). Intracranial and Abdominal Aortic Aneurysms: Similarities, Differences, and Need for a New Class of Computational Models. *Annual Review of Biomedical Engineering*, Vol. 10, pp. 221-246.

Inci, S. & Spetzler, R.F. (2000). Intracranial Aneurysms and Arterial Hypertension: A Review and Hypothesis. *Surgical Neurology*, Vol. 53, pp. 540-542.

Invesalius (2008). Invesalius Software: Basic Course. Cenpra, Campinas, São Paulo, Brazil.

Jou, L.D., Quick, C.M., Young, W.L., Lawton, M.T., Higashida, R., Martin, A. & Saloner, D. (2003). Computational Approach to Quantifying Hemodynamic Forces in Giant Cerebral Aneurysms. *AJNR American Journal of Neuroradiology*, Vol. 24, pp. 1804-1810.

Kaiser, D., Freyberg, M.A. & Friedl, P. (1997). Lack of Hemodynamic Forces Triggers Apoptosis in Vascular Endothelial Cells. *Biochemical and Biophysical Research Communications*, Vol. 24, pp. 586-590.

Kleinstreuer, C., Li, Z. & Farber, M.A. (2007). Fluid-Structure Interaction Analyses of Stented Abdominal Aortic Aneurysms. *Annual Review of Biomedical Engineering*, Vol. 9, pp. 169-204.

Kwak, D. & Kiris, C.C. (2011). *Computation of Viscous Incompressible Flows*, Springer, ISBN 978-94-007-0192-2, New York, USA.

Lakatta, E.G., Mitchell, J.H., Pomerance, A. & Rowe, G. (1987). Human Aging: Changes in Structure and Functions. *Journal of the American College of Cardiology*, Vol. 10, pp. 42A-47A.

Lasheras, J.C. (2007). The Biomechanics of Arterial Aneurysms. *Annual Review of Fluid Mechanics*, Vol. 39, pp. 293-319.

Lederle, F.A. & Simel, D.L. (1999). The Rational Clinical Examination. Does this Patient have Abdominal Aortic Aneurysm? *The Journal of the American Medical Association*, Vol. 281, pp. 77-82.

Legendre, D. (2010). Study of Flow Behavior Through Physical and Computational Infrarenal Aortic Aneurysm Model Obtained by CT Scan. Academic Thesis, University of São Paulo, São Paulo, Brazil.

Lerman, A. & Burnett, J.C. Jr. (1992). Intact and Altered Endothelium in Regulation of Vasomotion. *Circulation*, Vol. 86, pp. III12 – III9.

Li, Y.S., Haga, J.H. & Chien, S. (2005). Molecular Basis of the Effects of Shear Stress on Vascular Endothelial cells. *Journal of Biomechanics*, Vol. 38, pp. 1949-1971.

Long, Q., Xu, X.Y., Ariff, G., Thom, S.A., Hughes, A.D. & Stanton, A.V. (2000). Reconstruction of Blood Flow Patterns in a Human Carotid Bifurcation a Combined CFD and MRI Study. *Journal of Magnetic and Resonance Imaging*, Vol. 11, pp. 299-311.

MacSweeney, S.T.R., Powell, J.T. & Greenhalgh, R.M. (1994). Pathogenesis of Abdominal Aortic Aneurysm. *The British Journal of Surgery*, Vol. 81, pp. 935-941.

Malek, A.M., Alper, S.L. & Izumo, S. (1999). Hemodynamic shear stress and its role in atherosclerosis. *The Journal of the American Medical Association*, Vol. 282, pp. 2035-2042.

Masuda, H., Zhuang, Y.J., Singh, T.M., Kawamura, K., Murakami, M., Zarins, C.K. & Glagov, S. (1999). Adaptive Remodeling of Internal Elastic Lamina and Endothelial Lining During Flow-Induced Arterial Enlargement. *Arteriosclerosis, Thrombosis and Vascular Biology*, Vol. 19, pp. 2298-2307.

Meng, H., Wang, Z., Hoi, Y., Gao, L., Metaxa, E., Swartz, D.D. & Kolega, J. (2007). Complex Hemodynamics at the Apex of an Arterial Bifurcation Induces Vascular Remodeling Resembling Cerebral Aneurysm Initiation. *Stroke*, vol. 38, pp. 1924-1931.

Mesh, C.L., Baxter, B.T., Pearce, W.H., Chisholm, R.L., McGee, G.S. & Yao, J.S. (1992). Collagen and Elastin Gene Expression in Aortic Aneurysms. *Surgery*, Vol. 112, pp. 256-261.

Nichols, W.W. & O'Rourke, M.F. (1990). *McDonald's Blood Flow in Arteries: Theoretic, Experimental and Clinical Principles*. London: Edward Arnold.

Nichols, W.W., O'Rourke, M.F. (1990). *Properties of the Arterial Wall*. In McDonald's Blood Flow in Arteries. London: Lea & Febiger, pp. 81-85.

Paszkowiak, J.J. & Dardik, A. (2003). Arterial wall shear stress: Observations from the bench to the beside. *Vascular and Endovascular Surgery*, Vol. 37, pp. 47-57.

Patankar, S.V. (1975). *Numerical Prediction of Three-Dimensional Flows, in Studies in Convection: Theory Measurement and Applications*. Academic Press, London, Vol. 1.

Raaymakers, T. (1999). Aneurysms in Relatives of Patients with Subarachnoid Hemorrhage: Frequency and Risk Factors. MARS Study Group. Magnetic Resonance Angiography in Relatives of Patients with Subarachnoid Hemorrhage. *Neurology*, Vol. 53, pp. 982–988.

Rayz, V.L., Boussel, L., Acevedo-Bolton, G., Martin, A.J., Young, W.L., Lawton, M.T., Higashida, R. & Saloner, D. (2008). Numerical Simulations of Flow in Cerebral Aneurysms: Comparison of CFD Results and in vivo MRI Measurements. *Journal of Biomechanical Engineering*, Vol. 130, pp. 051011.

Reed, D., Reed, C., Stemmermann, G. & Hayashi, T. (1992). Are Aortic Aneurysms Caused by Atherosclerosis? *Circulation*, Vol. 85, pp. 205-21.

Ronkainen, A., Hernesniemi, J., Puranen, M., Niemitukia, L, Vanninen, R., Ryynanen, M., Kuivaniemi, H. & Tromp, G. (1997). Familial Intracranial Aneurysms. *Lancet*, Vol. 349, pp. 380–384.

Salsac, A.V., Sparks, S.R., Chomaz, J.M. & Lasheras, J.C. (2006). Evolution of the Wall Shear Stresses During the Progressive Enlargment of Symmetric Abdominal Aortic Aneurysms. *Journal of Fluid Mechanics*, Vol. 560, pp. 19-51.

Shah, P.K. (1997). Inflammation, Metalloproteinases, and Increase Proteolysis: an Emmerging Paradigm in Aortic Aneurysm. *Circulation*, Vol. 96, pp. 2115-2117.

Shojima, M., Oshima, M., Takagi, K., Torii, R., Hayakawa, M., Katada, K., Morita, A. & Kirino, T. (2004). Magnitude and Role of Wall Shear Stress on Cerebral Aneurysm: Computational Fluid Dynamic Study of 20 Middle Cerebral Artery Aneurysms. *Stroke*, Vol. 35, 2500-2505.

Shojima, M., Oshima, M., Takagi, K., Torii, R., Nagata, K., Shirouzu, I., Morita, A. & Kirino, T. (2005). Role of the Bloodstream Impacting Force and the Local Pressure Elevation in the Rupture of Cerebral Aneurysms. *Stroke*, Vol. 36, pp. 1933-1938.

Singh, K., Bonaa, K.H., Jacobsen, B.K., Bjork, L. & Solberg, S. (2001). Prevalence of and Risk Factors for Abdominal Aortic Aneurysms in a Population-Based Study. *American Journal of Epidemiology*, Vol. 154, pp. 236-244.

Tang, B.T., Cheng, C.P., Draney, M.T., Wilson, N.M., Tsao, P.S., Herfkens, R.J. & Taylor, C.A. (2006). Abdominal Aortic Hemodynamics in Young Healthy Adults at Rest and During Lower Limb Exercise: Quantification Using Image-Based Computer Modeling. *American Journal of Physiology. Heart and Circulatory Physiology*, Vol. 291, pp. H668-H676.

Taylor, C.A. & Figueroa, C.A. (2009). Patient-Specific Modeling of Cardiovascular Mechanics. *The Annual Review of Biomedical Engineering*, Vol. 11, pp. 109-134.

Thiriet, M. (2008). *Biology and Mechanics of Blood Flows: Part II – Mechanics and Medical Aspects*, Springer, ISBN 978-0-387-74848-1, New York, USA.

Thubrikar, M. J. (2007). *Vascular Mechanics and Pathology*, Springer, ISBN-10: 0-387-33816-0, New York, USA.

Toth, M., Nadasy, G.L., Nyary, I., Kerenyi, T., Orosz, M., Molnárka, G. & Monos, E. (1998). Sterically Inhomogenenous Viscoelastic Behavior of Human Saccular Cerebral Aneurysms. *Journal of Vascular Reserach*, vol. 35, pp. 345-355.

Utter, G. & Rossman, J.S. (2007). Numerical Simulation of Saccular Aneurysm Hemodynamics: Influence of Morphology on Rupture Risk. *Journal of Biomechancis*, Vol. 40, 2716-2722.

Vorp DA, Raghavan ML, Webster MW.(1998) Mechanical wall stress in abdominal aortic aneurysm: influence of diameter and asymmetry. *J Vasc Surg*;27:632-639.

Vorp, D.A. (2007). Biomechanics of Abdominal Aortic Aneurysm. *Journal of Biomechanics*, Vol. 40, pp. 1887-1902.

Wernig, F. & Xu, Q. (2002). Mechanical Stress-Induced Apoptosis in the Cardiovascular System. *Progress in Biophysisc & Molecular Biology*, Vol. 78, Issues 2-3, (February-April), pp. 105-137.

Zamir, M. (2005). *The Physics of Coronary Blood Flow*, Springer, ISBN 978-0387-25297-1, New York, USA.

Zhang, B., Fugleholm, K., Day, L.B., Ye, S., Weller, R.O. & Day, I.N. (2003). Molecular Pathogenesis of Subarachnoid Haemorrhage. *The International Journal of Biochemistry & Cell Biology*, Vol. 35, pp. 1341-1360.

7

Numerical Simulation for Intranasal Transport Phenomena

Takahisa Yamamoto[1], Seiichi Nakata[2], Tsutomu Nakashima[3]
and Tsuyoshi Yamamoto[4]

[1]*Gifu National College of Technology*
[2]*Fujita Health University*
[3]*Nagoya University*
[4]*Kyushu University*
Japan

1. Introduction

More than 10 million people in Japan suffer some of nasal diseases every year (Haruna (2003)); the paranasal sinusitis (so-called the empyema), hypertrophic rhinitis and inferior concha inflammation. Nebulizer treatment has been used for the nasal diseases. The effectiveness of the nebulizer treatment has been confirmed from clinical view points until now. However there are a few researches that evaluate the effect of the nebulizer treatment theoretically and quantitatively, i.e., the transport characteristics of medicinal droplets and their deposition on the inflammation areas of nasal wall.

The development of medical image processing technique is in progress and now gives us exquisitely detailed anatomic information. Some researchers calculated blood flow inside vital arteries and intranasal airflow characteristics by means of the medical image processing technique as well as Computed Fluid Dynamics (CFD) analysis. As for the CFD analysis of intranasal flow, Weinhold et al. constructed both a transparent resin model and numerical three-dimensional anatomy model with nasal cavities using a patient's CT data (Weinhold & Mlynski (2004)). They subsequently made clear airflow characteristics in the nose experimentally and numerically, and found that pressure drop was a main factor of nasal airflow. Lindemann et al. focused at a case which underwent radical sinus surgery (Lindemann et al. (2004; 2005)). In their case, both the lateral nasal wall and the turbinates, inhibiting physiological airflow, were removed by the surgery to realize the enlargements of the nasal cavity volumes and to increase the ratio between nasal cavity volume and surface area. However the researches mentioned here dealt with only a few patient-cases even though there are individual differences in the shape of human nasal cavity and in the grade of medical conditions. The past researches considered the individual differences insufficiently. The authors analyzed intranasal transport phenomena for several patient cases and compared each others in the past study (Monya et al. (2009); Yamamoto et al. (2009)). From the results characteristics of airflow and medicinal droplet transportation strongly depend on inflow conditions such as inflow angle, velocity and size of particle even if there are the individual differences for the shape of patient's nasal cavity.

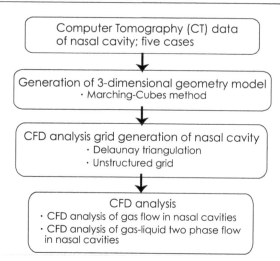

Fig. 1. Flowchart diagram of CFD analysis for heat and mass transport phenomena in nasal cavity

Figure 1 indicates the flowchart of the CFD analysis for intranasal transport phenomena. This chapter shows you detail features of nasal cavity and some nasal diseases, subsequently how to yield three-dimensional model of nasal cavity and calculation mesh for CFD analysis. Finally the characteristics of intranasal transport phenomena are presented in this chapter.

2. Anatomy of nasal cavity

2.1 Structure of nasal cavity and nasal sinuses
Figure 2 shows front and side views of human head, nasal cavity, nasal conchae and nasal sinuses. The nasal cavity is a large air filled space, and the pathway of respiration flow. The nasal cavity conditions the air to be received by the other areas of the respiratory tract. The nasal conchae, which has large surface area, warms and cools passing air through the nasal cavity. At the same time, the passing air is humidified and cleaned by nasal conchae and nasal hairs; most parts of dusts and particulate matters are removed there. The nasal sinuses are four caviums existing in the ossa faciei as shown in Fig.2; maxillary sinus, frontal sinus, ethmoidal sinus and sphenoidal sinus. These sinuses connect to the nasal cavity via very small ducts, ostia; both diameter and streamwise scales are several millimeters. Figures 3 represents coronal and sagittal cross-sectional CT data for nasal cavity and nasal sinuses. The white areas in the figure represent bones of the patient's head, and the gray areas indicate muscles and fat. The nasal cavity and sinuses are represented as dark area in the CT data.
In paranasal sinusitis, both the sinuses and the ostia are infected by bacillus and the virus. The details of such the diseases are stated in following subsections.

2.2 Nasal diseases
The nose often suffers some injuries such as fractures because the nose is protruded forward from the human face. Infection, epistaxis, as well as polyps are also found in the nasal diseases. Almost all people have contract rhinitis which is caused by the inflammation of nasal mucosa. Sinusitis will develop when the inflammation of the rhinitis spread to the sinuses via ostia.

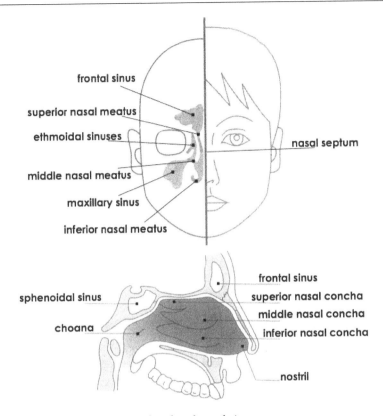

Fig. 2. Front and side view of human head and nasal sinuses (adapted from Haruna (2003))

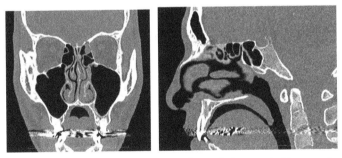

Fig. 3. Cross-sectional CT data for nasal cavity and nasal sinuses: lett) coronal and right) sagittal sections

2.2.1 Deviated nasal septum

Nasal septum locates on the middle of left and right nasal cavities, being almost straight stretches. Many people has a slightly-curved nasal septum as shown in the left-side figure of Fig.5. There are small size different between the left and right nostrils. The right figure of Fig.5 indicates a CT image that is the case of deviated nasal septum. The deviated nasal septum is caused by external nasal injuries and/or congenital origin. As for the cases, one of the nasal

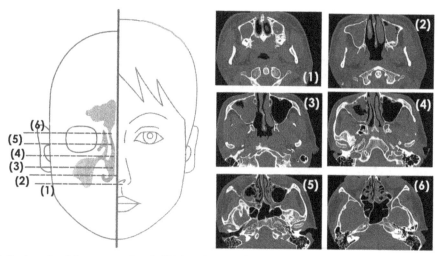

Fig. 4. Series of axial cross-sectional CT data for nasal cavity and nasal sinuses (adapted from Haruna (2003))

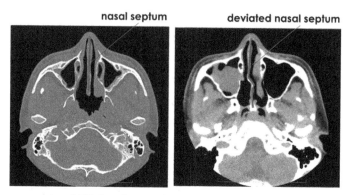

Fig. 5. Axial cross-sectional CT data for nasal septum: left) healthy nasal septum and right) deviated nasal septum

cavity is significantly smaller than the other. Then the patient suffers nasal congestion and drying nasal cavity, consequently epistaxis. In the sever cases operative treatment is required.

2.2.2 Rhinitis (nasal inflammation)

The nasal cavity is the part that easily causes infection in the upper respiratory tracts. Rhinitis (nasal inflammation) is mainly classified into two cases; the one is acute rhinitis and the other is chronic rhinitis. The acute rhinitis is generally caused by the infection of virus and allergies of pollen and house dust. The difference point between both rhinitis is the term of these diseases; the former occurs in a short term and the latter in a long term (two weeks and more). In some cases, the chronic rhinitis occasionally accompanies the chronic sinusitis.

2.2.3 Sinusitis

Figure 6 shows cross-sectional CT data for both healthy maxillary sinus and sinusitis one. The sinusitis happens anywhere about four kinds of the sinus paranasalis such as the maxillary sinus, sinus ethmoidales, the master wrestlers caves, and the sphenoidal sinuses. The sinusitis is classified into the acute sinusitis (in a short term) and the chronic sinusitis (in a long term).

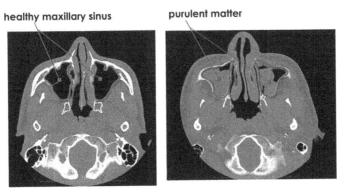

Fig. 6. Cross-sectional CT data for maxillary sinuses: left) healthy maxillary sinus and right) sinusitis

The acute sinusitis is caused by a variety of bacteria and virus, often develops after the blockage of ostia which are opening of sinuses connecting to nasal cavity. The blockage normally occures as a result of virus infection in the upper respiratory caused by a cold. The cold brings on the swelling of the mucous membrane of nostril, and then the ostia are easily obstructed. In the blocked sinus, the air inside the cavity are absorbed to the bloodstream, and subsequently the pressure inside the sinus decreases. This pressure drop produce sever pains and the sinus fills with secretory fluid. The secretory fluid becomes a breeding ground for viruses. In order to attack the viruses, white blood cell and other matters such as the secretory fluid are aggregated in the blocked sinus, then the pressure in the sinus is increased and pain becomes more sever.

The case that a symptom of the sinusitis continues more than 8-12 weeks is called the chronic sinusitis. The developing mechanism of the chronic sinusitis has not been clear, however it has been confirmed that the chronic sinusitis develops after the infection of the virus, severe allergies and the influences of the environmental pollution material. A genetic contributor is also regarded as one of the factors concerned with development of the chronic sinusitis.

3. Construction of three-dimensional nasal cavity model

In order to construct a three-dimensional model for nasal cavity, data conversion algorithm from two-dimensional CT/MRI data to three-dimensional geometric model as well as smoothing algorithm for the three-dimensional model are required. Furthermore, mesh generation models are needed to execute CFD analysis of intranasal transport phenomena. This section describes the fundamental theories of these algorithms.

3.1 Data conversion algorithm from 2-D to 3-D: marching cubes algorithm

Marching cubes algorithm is one of the latest models of surface construction used for viewing three-dimensional data (Lorensen & Cline (1987)). This algorithm extracts a polygonal mesh

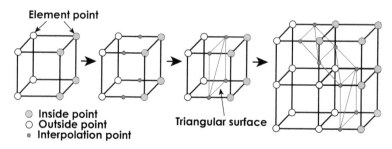

Fig. 7. Configurations of fourteen unique cubes for marching cubes algorithm

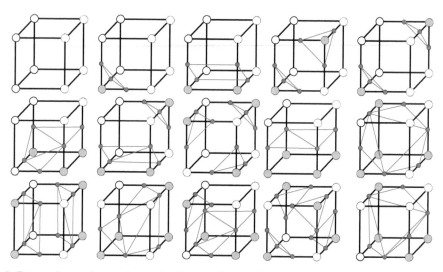

Fig. 8. Triangular surface patterns for the marching cubes algorithm

of isosurface from a three-dimensional scalar field. The algorithm proceeds through the scalar field, taking eight neighbor locations at a time, then determining the polygons needed to represent the part of the isosurface that passes through this cube. The individual polygons are then fused into the desired surface as shown in Fig.7. This is done by creating an index to a pre-calculated array of 256 possible polygon configurations ($2^8 = 256$) within the cube, by treating each of the 8 scalar values as a bit in an 8-bit integer. Finally each vertex of the generated polygons is placed on the appropriate position along the cube's edge by linearly interpolating the two scalar values that are connected by that edge. The precalculated array of 256 cube configurations can be obtained by reflections and symmetrical rotations of 14 unique cases as shown in Fig.8.

The applications of this algorithm are mainly concerned with medical visualizations such as CT and MRI scan data images, and special effects on three-dimensional modeling.

3.2 Smoothing algorithm

In the image processing and image transmission, noises are physically captured and fed into the processes due to various factors. For instance, noises are generated caused by the image resolution and image slice thickness of CT and MRI in the field of medical engineering. When the noise generation mechanism is clear and mathematically modeled, we can remove the noise using the optimized filter. However, it is difficult to modeled the noise mathematically in almost all situations.f The smoothing processes are needed to reduce and minimize the influence of the noise. The noise makes patterned indented surfaces in the three-dimensional model, therefore affects the quality of the constructed nasal cavity model significantly. There are two smoothing algorithms that often uses in the medical image processing; moving-average algorithm and median-average algorithm. This subsection presents both algorithms as follows.

3.2.1 Moving-average algorithm

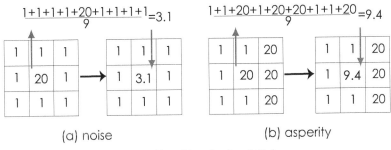

(a) noise   (b) asperity

Fig. 9. Procedure of Moving-Avarage Algorithm for 3×3 Cubes

Figure 9 shows the application of the moving-average algorithm for 3×3 cubes (Savitzky & Golay (1964); Sun (2006); Tamura (2002)). The number inside a cube expresses a scalar variable, for example the concentration of a chemical specie. The case of Fig.9(a) presents a noise included in the data and that of Fig.9(b) expresses an asperity in the data. The moving-average algorithm calculates an average of surroundings of the concentration $f(i,j)$ in the input image, and converts to the concentration $g(i,j)$ in the output image.

$$g(i,j) = \frac{1}{n^2} \sum_{k=-\lceil n/2 \rceil}^{n/2} \sum_{l=-\lceil l/2 \rceil}^{n/2} f(i+k,j+l) \tag{1}$$

The moving-average algorithm feathers the edge and the boundary of the input image, consequently smoothes the object surfaces.

3.2.2 Median-average algorithm

Figure 10 indicates the application of the median-average algorithm for 3×3 cubes (Boyle & Thomas (1988); Tamura (2002)). The cases of Fig.10(a) and (b) represent a noise and an asperity in the data, respectively. This algorithm sorts both a concentration value $f(i,j)$ at a target position of the input image and its surroundings in ascending order, subsequently converts the output image $g(i,j)$ using their median value. As for Fig.10(a), the concentration values in input image are rearranged like "$1,1,1,1,1,1,1,1,20$" in ascending order, and then the fifth value "1" is picked up as a median value and used as the concentration $g(i,j)$ in the

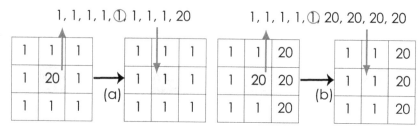

Fig. 10. Procedure of Median-Average algorithm for 3×3 cubes

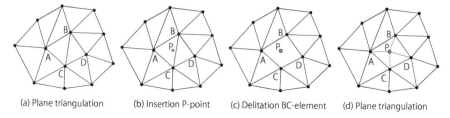

(a) Plane triangulation   (b) Insertion P-point   (c) Delitation BC-element   (d) Plane triangulation

Fig. 11. Schematic drawing of the Delaunay triangulation model

output image. On the other hand, the concentration values in input image are rearranged like "1,1,1,1,1,20,20,20,20" in ascending order as shown in Fig.10(b). Finally, the fifth value "1" is selected as the concentration $g(i, j)$ in the output image. The median-average algorithm can remove the noise effectively compared with the moving-average algorithm. At the same time, this algorithm involves a risk that the magnitude of rearranging the original CT/MRI data is larger than the moving-average algorithm, consequently could yield different three-dimensional model from the real target.

3.3 Mesh generation: Delaunay triangulation model
The Delaunay triangulation model is one of the most-used mesh generation technique. This model is applicable for a wide variety of very complicate geometries.
This model mathematically ensures that the circumcircle associated with each triangle contains no other point in its interior. This definition extends naturally to three-dimensional problems. Figure 11 shows a schematic drawing of the Delaunay triangulation model (Nakahashi (1995)). The triangles as shown in Fig.11(a) contains the edges and the edge's endpoints. A new point "P" is added inside the triangle "ABC", then the mathematical theory of the Delaunay triangulation is not satisfied in the triangle "ABC" and its surrounding triangle "BCD". The model can eliminate the edge "BC" which is a common edge for both triangles "ABC" and "BCD", subsequently reconstruct triangles using the added point "P". Consequently, as shown in Fig.11(d), the model can yield four triangles "ABP", "ACP", "BDP" and "CDP". The Delaunay triangulation constructs and reconstructs triangles in target domain simultaneously and give us fine meshes for CFD analysis.

4. Theory of computational fluid dynamics

In the nasal cavity, particulate matters such as pollens, house dust as well as medicinal droplets are transported via airflow of the respiration and the nebulizer treatment. In order to execute CFD analysis for such the transport phenomena, we have to consider multi-phase flow. This section describes the basic theory of CFD analysis for two-phase flow.

4.1 Governing equations for continuous phase

The Reynolds number of intranasal flow is a few hundred. Therefore the flow is assumed as incompressible and laminar flow.
The equation for conservation of mass, or continuity equation, can be written as follows:

$$\frac{\partial \rho_F}{\partial t} + \nabla \cdot \rho_F \mathbf{u_F} = 0 \tag{2}$$

here ρ_F and $\mathbf{u_F}$ are density and velocity of continuous phase. Equation 2 is the general form of the mass conservation equation and is valid for incompressible as well as compressible flows. Conservation of momentum in an inertial reference frame is described by

$$\frac{\partial \rho_F \mathbf{u_F}}{\partial t} + \mathbf{u_F} \cdot \nabla \rho_F \mathbf{u_F} = -\nabla p + \mu_F \nabla^2 \mathbf{u_F} + \rho_F \mathbf{g} + \mathbf{F} \tag{3}$$

where p is the static pressure, μ_F is viscosity of the fluid, and $\rho \mathbf{g}$ and $\mathbf{F}$ are gravitational body force and external body forces, respectively. $\mathbf{F}$ contains interaction with the dispersed phase in multi-phase flow and other model-dependent source terms.
The thermal energy transport equation is meant to be used for flows which are low speed and close to constant density.

$$\frac{\partial \rho_F c_p T_F}{\partial t} + \mathbf{u_F} \cdot \nabla \rho_F c_p T_F = \nabla \cdot (\lambda_F \nabla T_F) + S_{FP} \tag{4}$$

where c_p and λ_F are specific heat at constant pressure and heat conductivity, respectively. S_{FP} denotes a source term regarding the interaction between continuous phase and disperse phase.

4.2 Governing equations for disperse phase

Consider a discrete particle traveling in a continuous fluid medium. The forces acting on the particle which affect the particle acceleration are due to the difference in velocity between the particle and fluid, as well as to the displacement of the fluid by the particle. The equation of motion for such a particle was derived by Basset, Boussinesq and Oseen for a rotating reference frame (Akiyama (2002)):

$$m_p^i \frac{d \mathbf{u_P}}{dt} = \mathbf{F}_{Drag}^i + \mathbf{F}_{Pressure}^i + \mathbf{F}_{Basset}^i + \mathbf{F}_{Buoyancy}^i + \mathbf{F}_{Rotation}^i + \mathbf{F}_{Others}^i \tag{5}$$

where, superscript i denote the i-th particle, and $\mathbf{F}_{Drag}$, $\mathbf{F}_{Pressure}$, $\mathbf{F}_{Basset}$, $\mathbf{F}_{Buoyancy}$, $\mathbf{F}_{Rotation}$ express drag force acting on the particle, pressure gradient force, Basset force, buoyancy force due to gravity and forces due to domain rotation, respectively. The pressure gradient force term applies on the particle due to the pressure gradient in the fluid surrounding the particle caused by fluid acceleration. The Basset force term accounts for the deviation in flow pattern from a steady state.
The aerodynamic drag force on a particle is propotional to the slip velocity between the particle and the fluid velocity

$$\mathbf{F}_{Drag}^i = \frac{1}{8} \pi (d^i)^2 \rho_F C_D \mathbf{u_F} - \mathbf{u_P} (\mathbf{u_F} - \mathbf{u_P^i}) \tag{6}$$

where $\mathbf{u_F}$ and $\mathbf{u_P}$ are the velocities of the fluid flow and the particle. C_D denotes the drag coefficient of the particle.

$$C_D = \frac{24}{Re^i(1 + 0.15(Re^i)^{0.687})} \tag{7}$$

Re^i is the Reynolds number based on the slip velocity between the fluid flow and the i-th particle velocities.

$$Re^i = \frac{d^i(\mathbf{u_F} - \mathbf{u_P^i})}{\nu_F} \tag{8}$$

The pressure gradient force results from the local fluid pressure gradient around the particle and is defined as

$$\mathbf{F_{pressure}^i} = -\frac{1}{6}\pi(d^i)^3 \nabla p \tag{9}$$

This force is only important if large fluid pressure gradients exist and if the particle density is smaller than or similar to the fluid density.

$$\mathbf{F_{Basset}^i} = \frac{3}{2}(d^i)^2 \sqrt{\pi \rho_F \mu_F} \int_{t_0}^{t} \frac{d(\mathbf{u_F} - \mathbf{u_P^i})}{d\tau}(t-\tau)^{-\frac{1}{2}} d\tau \tag{10}$$

The buoyancy force is the force on a particle immersed in a fluid. The buoyancy force is equal to the weight of the displaced fluid and is given by

$$\mathbf{F_{Buoyancy}^i} = \frac{3}{2}(d^i)^3(\rho_P^i - \rho_F)\mathbf{g} \tag{11}$$

here $\mathbf{g}$ expresses the gravity vector.

In rotating frame of reference, the rotation term is an intrinsic part of the acceleration in and is the sum ob Coriolis and centripetal forces.

$$\mathbf{F_{rotation}} = -\frac{1}{3}(d^i)^3 \rho_P(\omega \times \mathbf{u_P}) - \frac{1}{6}\rho_P\omega \times (\omega \times \mathbf{r_P}) \tag{12}$$

where, ω denotes rotation speed of the reference.

In the intranasal flow and heat and mass transfer, the influence of both the pressure gradient term and the Basset force term on the transport phenomena in the nasal cavity is extremely small. The momentum conservation equation for the particle is shown as

$$m_P^i \frac{d\mathbf{u_P^i}}{dt} = \frac{1}{8}\pi(d^i)^2 \rho_F C_D \mathbf{u_F} - \mathbf{u_P^i}(\mathbf{u_F} - \mathbf{u_P^i}) \tag{13}$$

When the Lagrangian method is applied to the particle movement, the location of the particle is given by

$$\frac{d\mathbf{x_P^i}}{dt} = \mathbf{u_P^i} \tag{14}$$

As for the almost all particulate matter transported in nasal cavity, the mass fraction of the disperse phase are extremely small compared with the continuous phase. Therefore it is only necessary to consider the one-way interaction between both phases, namely only the continuous phase influences the disperse phase.

Heat transfer concerning the disperse phase is governed by three physical processes, namely convective heat transfer, latent heat transfer associated with mass transfer, and radiative heat transfer. Since the heat transfer of intranasal flow is not so high temperature condition, we

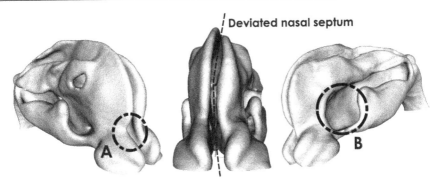

Deviated nasal septum

A B

Fig. 12. Three-dimensional nasal cavity model (case-1)

Case No.	Deviated nasal septum	Chronic sinusitis	hypertrophic rhinitis	Bloating inferior concha
1	○	Left	Left	
2		Left & Right		
3	○		Left & Right	Left
4	○		Left & Right	
5	○		Left & Right	Left

Table 1. Detail patient's case data

can neglect both the latent heat transfer and the radiative heat transfer. The convective heat transfer Q_C is given by

$$Q_C^i = \pi d^i \lambda_F Nu(T_F - T_P^i) \tag{15}$$

where T_P^i is the temperature of the i-th particle, and Nu is the Nusselt number given by

$$Nu = 2 + 0.6Re^{0.5}(\mu_F \frac{C_p}{\lambda_F})^{\frac{1}{3}} \tag{16}$$

The convective heat transfer Q_C is calculated and used in the source term of the thermal energy transport equation, Eq.4.

5. Numerical simulations of intranasal transport phenomena

5.1 Case data and three-dimensional model

Three-dimensional models of nasal cavity were constructed by means of five actual patient's CT data. These cases suffer several grades of deviated nasal septum symptom, hypertrophic rhinitis, chronic sinusitis and inferior nasal concha swelling. The case data are summarized in Table 1. Since head CT data involves not only nasal cavity but also skin, fat, bone and purulent matter, it is required to eliminate them. This study used three-dimensional medical image processing software "Mimics (Materialize Inc.)" and the marching cubes algorithm to determine the surfaces of nasal wall. The moving-average algorithm is also applied to smoothing the surface. Figure 12 shows the three-dimensional nasal model of case-I constructed in this study. The anterior and the posterior of the figure show anterior nares and upper pharynx, respectively. The three-dimension model has curved nasal septum which corresponds to the curvature as shown in Fig.12. The cross-sectional areas of nasal cavities

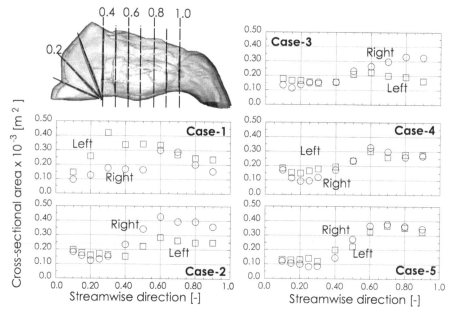

Fig. 13. Summaries of cross-sectional areas of nasal cavities for all cases

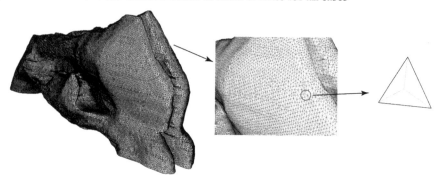

Fig. 14. Schematic drawing of CFD analysis grid (unstructured grid)

which significantly affect intranasal flow are summarized in Fig.13. The lumen and the wall of three-dimensional models are meshed using commercial meshing software ICEM CFD (ANSYS, Inc.) and the delaunay triangulation algorithm as shown in Fig.14.

Figure 15 indicates a example of the influence of the mesh quality on the accuracy of CFD analysis. In this figure, pressure drop between inlet and outlet and the number of grid nodes are used as the factors of the accuracy and the mesh quality, respectively. The pressure drop significantly depends on the number of grid nodes; the pressure drop shows occasional ups and downs in the case of small number of grid nodes, on the contrary it becomes stable in the case of large number of grid nodes.

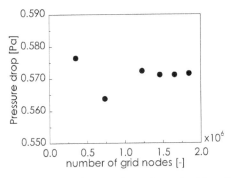

Fig. 15. Dependency between pressure drop and number of grid nodes in the CFD analysis for case-I

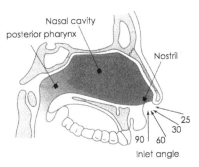

Fig. 16. Schematic figure of inlet nebulizer angles; 90, 60 and 30 degree

5.2 Computational fluid dynamics model

This study used CFD analysis code "CFX-11 (ANSYS, Inc.)" to analyze the intranasal transport phenomena of the nebulizer treatment. This code is able to solve partial differential equations for mass, momentum and energy transportations with appropriate boundary conditions. Since the Reynolds number of intranasal flow was several hundred, normal laminar flow analysis was used in this study. The Lagrange approach is adopted to solve transportation of the droplets of medicinal mist. Then the model employs assumptions that a droplet is sensitive with fluid flow, accelerating by velocity difference between fluid flow and the droplet. Air inlet velocity is set at 0.5m/s. This velocity condition corresponds to the condition of actual nebulizer treatment. As shown in Fig.16, four air inlet angles are considered in this study; 90, 60, 30 and 25 degree. In the nebulizer treatment, medicinal mist is atomized and transported by dry air. This study considers dry air (298K, 0.1MPa) as a fluid medium. No-slip condition, where fluid velocity on the wall surface is set at zero, is adopted to the nasal wall. Droplet diameter of medicinal mist is uniformly 50 micro meters. This value also corresponds to the condition of actual nebulizer treatment.

5.3 Results and discussion

Figures 17-19 shows stream lines in case-I left nasal when inlet velocity is 0.5m/s and nebulizer angles are 90, 60 and 30 degree. As we can see from the stream lines, changing nebulizer angle 60 to 30 degree changes the main flow of nebulizer from the superior nasal concha (upper-area of nasal cavity) to the inferior nasal concha (lower-area). In the all

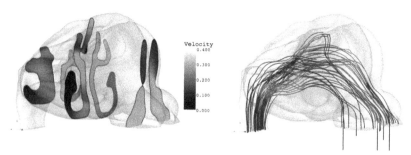

Fig. 17. Results of intranasal flow of the case-I; velocity magnitude distribution in each cross-section and streamlines (inflow velocity of 0.5m/s and inlet angle of 90 degree)

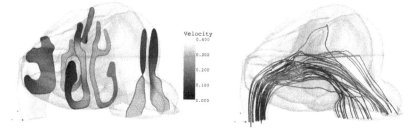

Fig. 18. Results of intranasal flow of the case-I; velocity magnitude distribution in each cross-section and streamlines (inflow velocity of 0.5m/s and inlet angle of 60 degree)

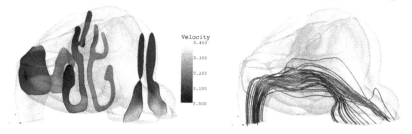

Fig. 19. Results of intranasal flow of the case-I; velocity magnitude distribution in each cross-section and streamlines (inflow velocity of 0.5m/s and inlet angle of 30 degree)

nebulizer angle conditions flow velocity increases near the nares. Especially, the increasing flow velocity is most obvious in the inflammation regions. Circulation flow is found in the lower inlet angle conditions. This circulation flow will be able to extend residence times of airflow and medicinal droplet in nasal cavity, consequently enhance the effect of the nebulizer treatment.

Figure 20-22 shows trajectories of medicinal droplets is the case-III when inlet velocity is 0.5m/s and inlet angles are 90, 60 and 30 degree. Almost all droplets impinge and deposit near the nares and middle nasal concha where the cross-section areas are narrow caused by deviated nasal septum and inflammation. From these results, the nebulizer angle is able to control the impingement and deposition characterisitics of medicinal droplets. By the control

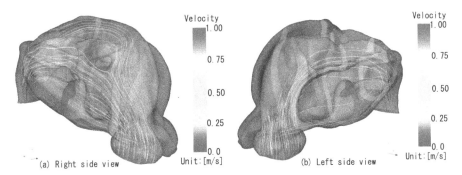

Fig. 20. Trajectories of nebulizer droplets of the case-III; inflow velocity of 0.5m/s and inlet angle of 90 degree)

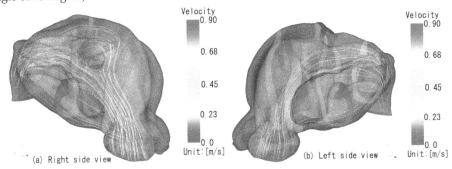

Fig. 21. Trajectories of nebulizer droplets of the case-III; inflow velocity of 0.5m/s and inlet angle of 60 degree)

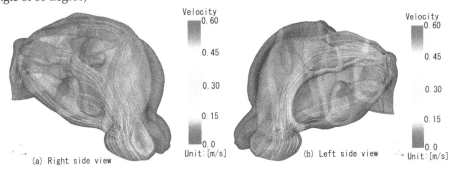

Fig. 22. Trajectories of nebulizer droplets of the case-III; inflow velocity of 0.5m/s and inlet angle of 30 degree)

of nebulizer angle, we can reduce the amount of medicinal droplet's deposition on nares and middle nasal concha, consequently transport wide area of nasal cavity.

This study focuses and uses three parameters to organize the results of each case; pressure drop of nebulizer flow, maximum velocity of intranasal flow and deposition ratio which expresses how many droplets impinge and deposit on nasal wall.   Figures 23-25 are correlations among them for all patient's cases. The correlations of each case show similar

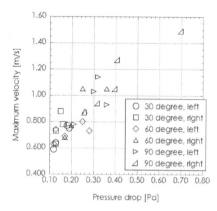

Fig. 23. Correlation analysis between maximum velocity and pressure drop

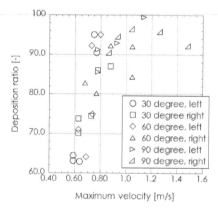

Fig. 24. Correlation analysis between deposition ratio and maximum velocity

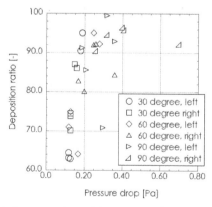

Fig. 25. Correlation analysis between deposition ratio and pressure drop

tendencies even though there are the individual differences such as the shape of nasal cavity and the state of nasal disease. Therefore these parameters are useful to organize the results mentioned above. The maximum velocity has direct proportional relationship with pressure drop. Meanwhile pressure drop and maximum velocity have strong correlation with deposition ratio. These results lead us to conclude that the nebulizer conditions have same-level influence on the intranasal transport phenomena for all cases.

6. Conclusion

This chapter presents not only the procedure of CFD analysis for intranasal transport phenomena but also medicinal droplets transport characteristic in patient's nasal cavity using actual CT/MRI data. Furthermore, the result of correlation analysis for the nebulizer treatment condition is also presented here. As for the results of CFD analysis, nebulizer angle, inlet velocity and size of droplet significantly affect on the intranasal transport phenomena. The condition of low nebulizer angle achieves that the medicinal droplets are transported all over the nasal cavity, that is this condition will increases therapeutic response to the nasal diseases. The correlation among velocity of intranasal flow, pressure drop and deposition ratio, which are main parameters for intranasal transportation of medicinal droplets in the nebulizer treatment, indicates that the basis of the intranasal transportation show similar characteristics in all patient's cases. These results mean that there is an optimum condition of nebulizer treatment for all patients who suffer nasal diseases. Further research of the intranasal transport phenomena for many patient's cases is required to survey the best treatment condition.

Since the development of the medical image processing technique has been in progress, the CFD analysis for human body, such as blood flow from the heart, respiration flow in the pharynx and the lung, will be widely spread and achieve tangible results in clinical practices within the next several years.

7. Acknowledgement

This work was supported by the grant from JSPS KAKENHI 19760114.

8. References

Akiyama, M. (2002). *Analysis of Two-Phase Flow Dynamics: Multi-Physics Flow Analysis*, Korona Inc., Tokyo.

Boyle, R. & Thomas, R. (1988). *Computer Vision: A First Course*, Blackwell Scientific Publications, New York.

Haruna, S. (2003). *An endoscope surgical operation of chronic sinus infection*, Hokendojinsha Inc., Tokyo.

Lindemann, J., Keck, T., Wiesmiller, K., Sander, B., Brambs, H.-J., Rettinger, G. & Pless, D. (2004). A numerical simulation of intranasal air temperature during inspiration, *The Laryngoscope* Vol.114(No.1): 1037–1041.

Lindemann, J., Keck, T., Wiesmiller, K., Sander, B., Brambs, H.-J., Rettinger, G. & Pless, D. (2005). Numerical simulation of intranasal airflow after radical sinus surgery, *American Journal of Otolaryngology* Vol.26(No.1): 175–180.

Lorensen, W. E. & Cline, H. E. (1987). Marching cubes:, *Computer Graphics: A high Resolution 3D Surface Construction Algorithm* Vol.21(No.4): 163–169.

Monya, M., Yamamoto, T., Nakata, S., Nakashima, T., Yamamoto, T. & Suzuki, T. (2009). Cfd analysis for intranasal mass transportation in deviated septum case (japanese), *JSME Journal Series B* Vol.75(No.751): 175–180.

Nakahashi, K. (1995). *Grid Generation and Computer Graphics*, Univ. of Tokyo, Tokyo.

Savitzky, A. & Golay, M. J. E. (1964). Smoothing and differentiation of data by simplified least squares procedures, *Anal. Chem.* Vol.36(No.8): 1627–1639.

Sun, C. (2006). Moving average algrithms for diamond, hexagon, and general polygonal shaped window operation, *Pattern Recognition Letters* Vol.27(No.6): 1627–1639.

Tamura, H. (2002). *Computer Image Processing*, Ohm Inc., Tokyo.

Weinhold, I. & Mlynski, G. (2004). Numerical simulation of air flow in the human nose, *Eur Arch Otorhinolaryngol* Vol.261(No.1): 452–455.

Yamamoto, T., Nakata, S., Nakashima, T. & Yamamoto, T. (2009). Computational fluid dynamics simulation for intranasal flow (japanese), *Foresight of Otorhinolaryngology* Vol.52(No.1): 24–29.

Part 2

Additional Important Themes

8

Fluid Dynamics Without Fluids

Marco Marcon

Politecnico di Milano, Dipartimento di Elettronica e Informazione,
Milano
Italy

1. Introduction

This chapter will discuss some interesting real applications where Fluid Dynamics equations found fruitful applications without dealing with "strictly speaking" fluids. In particular, thanks to the large set of analyses performed over different kinds of fluids in different operating and boundary conditions, a wide range of Computational Fluid Dynamics algorithms flourished tackling different aspects, from convergence rate, to stability according to the discretization, to multigrid and linearization problems. This robust and thorough background, both on theoretical and on practical aspects, made Computational Fluid Dynamics (CFD) appealing also to other sciences and applications where Fluid Dynamics equations, or similar equations very close to them, can be useful in describing complex phenomena not related to fluids. Some applications that will be discussed concern, e.g., Geometry of liquid snowflakes whose contour is growing steered by curvature, staring from a circle. Furthermore Image Restoration and Segmentation can also benefit from CFD since a set of evolutionary algorithms, based on level-set curvature flow equations, plays a fundamental role in steering active contours or snakes through the noise present in the image till the complete warping of the desired framed object. Also in this case advanced techniques like Ghost Fluids Method for two competing fluids dynamics can be used to separate different objects in images. Other interesting applications that will be described concern applicability of CFD to surface extraction from cloud of points. This is a common problem when complex clouds of points, representing 3D objects or scenes are obtained by laser scanners or multi-camera vision systems. These points represent unambiguous features from corners or sharp edges and the final 3D closed surface must fit on these points smoothly interpolating empty space between them. Also in this case CFD can provide useful tools to define the evolution of a 3D surface representing the border between two competing fluids, one representing the "inside" and the other the "outside" of the object itself. The two fluids evolution will stop when surface sticks on all the 3D points: the viscosity of the two fluids will control the smoothness of this surface that will wrap the cloud and turbulence is used to model injection into grooves or narrow holes. This chapter will also discuss another interesting application of CFD to robotic navigation in complex environments where we are looking for the best path, both in terms of length and distance from objects, through a set of obstacles, different terrains traversability or path slope. Also in this case an imaginary fluid with a predefined viscosity floods from the robot position through the whole environment, its front evolution speed, accordingly to CFD, will be slower in narrow passages and, once it reaches the target, it will define the easiest way.

2. Snowflakes and phase interfaces

In many physical problems outlines of a considered fluid have different speeds at different points, and in many cases local speed is directly connected to the curvature. In the following we will show how a geometrical contour, represented by a 2D line or a 3D surface of the evolving shape changes from point to point accordingly to the local geometrical properties of the contour itself. The presented approach can be applied to a generic N-dimensional case and we can refer to the moving contour as a general hypersurface, anyway in the following, we will limit without loss of generalization our considerations to a 3D space where fluid boundary is represented by a closed surface.

In particular Sethian (1989) showed how algorithms based on direct parametrization of the surface evolution could present ambiguous and error-prone solutions due to the local error propagation while global approaches based on implicit representations can result in a much more robust solution. This kind of approaches usually relies on higher dimensional functions for which the considered surface is just represented by a level-set. The contour motion can then be described applying the proper Hamilton-Jacobi equation related to a hyperbolic conservation law to the implicit function and tracking the particular level-set representing the contour itself.

Implicit function formulation is much more robust to numerical techniques with respect to direct methods and can be implemented starting from an original closed and non-intersecting surface: two famous physical examples are the combustion model of a flame and the grow of a snowflake: Markstein (1951) assumed that thin flame fingers close to their ends are cooler than their inner part and move slower that hot nonconvex regions: the proposed evolution equation assumes that flame contour speed is inversely proportional to its curvature κ, (where κ is positive for convex regions and negative for nonconvex ones). This assumption come from the fact that in highly convex regions, flame particles collide with a high number of slower and cooler surrounding particles that slow down their motion. The symmetrical effect is present in spikes growing in snowflakes. In particular these two last examples, flames and snowflakes, represent two very good examples where simple equations can explain very well complex phenomena. For a comprehensive description of ice crystal growth we suggest Libbrecht (2005), while in the following we describe how solid-liquid interface dynamics are steered by curvature in liquid snowflakes (or Tyndall figures). We refer to some recent studies (Hennessy (2010), Hobbs (2010)) on cylindrical discs of liquid in superheated crystals of ice. In this case the snowflake is made of water inside ice and its geometrical shape evolution is determined by boundary condition between solid and liquid phase. In Fig. 1 there is a typical snapshot of a Liquid snowflake.

2.1 Liquid snowflakes surface evolution

The typical modelization of snowflakes surface growing is the Gibbs-Thomson equation where the interface temperature T_I, between the solid and the liquid part is ruled by the equation:

$$T_I(\kappa) = T_M \left(1 - \kappa \frac{\gamma}{\rho L}\right) \tag{1}$$

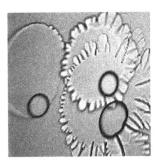

Fig. 1. A typical Tyndall figure (Liquid snowflake)

Where T_M is the equilibrium temperature of a planar solid-liquid interface ($273.15K$ for water), κ is the curvature, γ is the surface energy per unit area for the solid-liquid interface, ρ is the density of a specific phase but in this case we are assuming equal density both for ice and water and L is the latent heat of fusion. The steering interface force is then due to the pressure variation across the interface and depends from curvature:

$$p_s - p_l = \rho \left(s_l - s_s\right) \left(T_I - T_M\right) = \rho L \left(\frac{T_I \left(\kappa\right)}{T_M} - 1\right) \tag{2}$$

Where the subscript l and s refer to liquid and solid state respectively and s indicates the entropy per unit mass, for further details refer to (Hills & Roberts (1993)).

A further improvement in the Gibbs-Thomson model considers surface energy anisotropy due to crystal ice structure (six fold symmetry), this will add a dependence of the surface energy γ from the relative orientation of the surface with respect to the underlying crystal. Accordingly to (Davis (2001)) a six fold symmetrical crystal will have a:

$$\gamma_6 = \gamma_0 \left[1 + \frac{\sigma_n}{n^2 - 1} \cos\left(n\theta\right)\right]\Bigg|_{n=6} \tag{3}$$

Where σ_n represents the anisotropy degree and the complete interface temperature will be, accordingly to (Pimpinelli & Villain (1999));

$$T_I\left(\kappa\right) = T_M \left(1 - \kappa \frac{\gamma_6 + \frac{d^2\gamma_6}{d\theta^2}}{\rho L}\right) \tag{4}$$

In particular, accordingly to equations 3 and 4, whenever $|\sigma_n| < 1$ we are in a week anisotropy condition, i.e. the surface energy γ is always positive, and, at equilibrium, the surface will resemble a smooth line, while, for anisotropy values greater that one, as for the snowflakes water example, the thermodynamic equilibrium implies a closed polygonal made of straight line segments (Dobrushin et al. (1992)).

Once the Temperature of the interface is determined the shape of the boundary between the liquid and solid part and its time evolution can be recovered from the energy conservation, in particular, in order for the ice-water interface to advance inside the solid part, two energy contributions must be provided: one for solid melting and one for interface stretching. The melting energy that must be provided per unit of time for a surface A that is moving inside the solid at a normal speed v_n is $l_{mel} = \rho L A v_n$ while the energy contribution per unit of time due to surface stretching is $l_{sur} = \gamma \frac{dA}{dt}$; which, accordingly to the differential geometry can be written as:

$$l_{sur} = \gamma \frac{dA}{dt} = A \gamma \kappa v_n \tag{5}$$

The surface advancement is steered by two opposite heat diffusive fluxes: one inside the liquid part and one inside the solid part; respectively: $-k_l \nabla T_l$ and $k_s \nabla T_s$ where k indicates the thermal conductivity of the considered phase.

The equilibrium of energy equations per area and time unit can be written as:

$$\frac{l_{mel}}{A} + \frac{l_{sur}}{A} = (-k_l \nabla T_l + k_s \nabla T_s) \cdot \mathbf{n} \tag{6}$$

i.e.:

$$(\rho L + \gamma \kappa) v_n = (-k_l \nabla T_l + k_s \nabla T_s) \cdot \mathbf{n} \tag{7}$$

Where $\mathbf{n}$ is surface normal oriented from the liquid part towards the solid one; this equation is called the Stefan condition. The numerical solution for the liquid-solid interface evolution, accordingly to the Stefan condition, requires some particular precaution due to accuracy that must be placed in surface tracking. Since the analyzed problem is far from thermodynamic equilibrium due to the superheated temperature of the solid, some typical approaches, e.g. the Enthalpy model (Shastri & Allen (1998)), are not accurate. In fact, Enthalpy approach, even if requires only the evaluation of a single parabolic PDE at each time step, assumes that the interface is within a mushy solid-liquid region and the evolving interface is just approximated. A possible extension of Enthalpy approach to account for superheated solid state could be phase-field models; anyway an accurate interface tracking requires a very small grid spacing resulting in a high computational cost (Biben et al. (2005)). One of the method that in recent years is getting growing attention is Level-set: in the following of this chapter we will show it versatility and accuracy in many CFD applications and in liquid snowflakes shape definition.

3. Level-set methods

For most of real CFD systems symbolic formulation that easily yield to symbolic mathematical solution is unfeasible, and the goal becomes to find the system modelization with the best trade-off between accuracy and computational complexity accounting for local, global and independent properties of the moving surface. In particular we will refer to "local" as properties associated to local surface features (e.g. curvature, local density or temperature), "global" as properties associated to overall surface shape or extension, while independent properties are not linked to the surface itself, like a flow below the surface. The Level-set method aims to track a thin propagating interface over time, it outperforms many other approaches handling topological complexities such as corners and cusps, and in handling complexities in the evolving interface such as entropy conditions and weak solutions. A detailed description of level-set can be found in (Sethian (1999)).

The general idea is to define a function $\phi(p, t)$ over the whole space-time domain, where p indicates a point in the space and t represents the time variable. The initial surface is the zero level-set of an implicit function for $t = 0$: i.e. the starting surface is the set of points p so that $\phi(p, t) = 0$ the value of the ϕ function for non-interface points is, the signed distance from the surface where 'signed' means that different signs indicate internal or external position with respect to the closed surface. The discrete formulation requires that the value of ϕ is initially evaluated over all the mesh points and the steering equations would require the update of each point at each time step. This computational complexity of order $O(n^3)$ at each time step could be unacceptable in many practical cases so many faster update solutions have been proposed; the most common are:

- Narrow band: the updated mesh points are only those in a small stripe surrounding the interface, this will reduce computation to $O(n^2)$ but as soon as the interface is approaching the stripe boundary all values must be recomputed (more details in (Adalsteinsson & Sethain (1995))

- Octree: relevant mesh points and high detailed surface regions can be divided in sub-grids, e.g. a cube can be split in eight sub-cubes (the method name come from here) and the procedure can be iterated for higher precision. More details can be found in (Losasso et al. (2004))

- Sparse block grid: The whole domain is subdivided in blocks representing clusters of the original grid points (they can be overlapped or not) then the update is evaluated only on blocks containing the surface narrow band. (Further details in (Bridson (2003))

3.1 Level-set evolution

Accordingly to the previous description the set of points belonging to the surface will satisfy the zero level-set condition: $\phi(\mathbf{p}(t), t) = 0$ where $\mathbf{p}(t)$ represents the path of a surface point. The first temporal derivative will become:

$$\frac{d\phi(\mathbf{p}(t), t)}{dt} = \frac{\partial \phi}{\partial t} + \nabla \phi \cdot \frac{d\mathbf{p}(t)}{dt} = 0 \tag{8}$$

Considering the normal unitary vector to a generic point of ϕ for a generic level-set (contour), $\mathbf{n} = \frac{\nabla \phi}{|\nabla \phi|}$ the speed function v along this direction is then:

$$v(t) = \frac{d\mathbf{p}(t)}{dt} \cdot \mathbf{n} \tag{9}$$

obtaining:

$$\frac{d\phi(\mathbf{p}(t), t)}{dt} = \frac{\partial \phi}{\partial t} + v(t) |\nabla \phi| = 0 \tag{10}$$

In the cases considered in the following and in many physical cases the interface evolution is ruled by a Hamilton-Jacobi equation and the local velocity can be assumed proportional to the local curvature of the surface itself:

$$v = v_0 - \epsilon \kappa \tag{11}$$

where ϵ is a constant and

$$\kappa = \frac{\phi_{xx}\phi_y^2 - 2\phi_y\phi_x\phi_{xy} + \phi_{yy}\phi_x^2}{\left(\phi_x^2 + \phi_y^2\right)^{3/2}} \tag{12}$$

We can rewrite ϕ as follows:

$$\frac{\partial \phi}{\partial t} = -v_0 |\nabla \phi| + \epsilon \kappa |\nabla \phi| \tag{13}$$

3.2 Liquid snowflakes surface evolution

Assuming that the ice is radiated with an external electromagnetic radiation normal to its surface (a focused beam with a gaussian shape) the heat will follow an exponential decay penetration, accordingly to the following heat propagation equation:

$$I(r,z) = I_0 e^{-(r/r_b)^2 - \alpha z} \tag{14}$$

where I_0 is the central beam intensity, r_b is its standard deviation while r is the in-plane coordinate for the ice surface. Furthermore z is the coordinate orthogonal to the ice surface and α is a penetration coefficient. The formation of the liquid snowflake is then in the in-plane direction. A level-set function can the be defined where ϕ is the normalized temperature that follows heat PDE:

$$\begin{cases} \frac{\partial \phi}{\partial t} = \nabla^2 \phi + e^{-(r/\beta)^2} & \text{for the liquid} \\ c_{pr} \frac{\partial \phi}{\partial t} = k_r \nabla^2 \phi + \alpha_r e^{-(r/\beta)^2} & \text{for the solid} \end{cases} \tag{15}$$

where $c_{pr} = \frac{c_{ps}}{c_{pl}}$ is the relative measure of heat capacities, $k_r = \frac{k_s}{k_l}$ is the relative thermal conductivity and $\alpha_r = \frac{\alpha_s}{\alpha_l}$ is the relative absorbtion coefficient.

The liquid/solid interface is then associated to the zero level-set in a sort of rescaled Celsius scale. Experimental observations show that liquid snow flakes often begin as circular discs which then grow outwards in a radially symmetric manner. However, after a certain amount of time, the interface becomes unstable and small, sinusoidal perturbations with a well defined wave-number appear. In Fig. 2 it is possible to see the growing of a liquid snowflake starting from a circle and then evolving with equally spaced rippling.

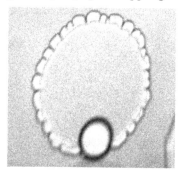

Fig. 2. Liquid snow flakes formation starting from a circle and the evolving, accordingly to linear instabilities with sinusoidal perturbations

The eq. 15, accordingly to the problem symmetry can be written in polar coordinates for the quasi-steadily condition:

$$\begin{cases} \frac{1}{r} \frac{\partial}{\partial r} \left(r \frac{\partial \phi}{\partial r} \right) + \frac{1}{r^2} \frac{\partial^2 \phi}{\partial \theta^2} + e^{-(r/\beta)^2} = 0 & \text{for the liquid} \\ \frac{1}{r} \frac{\partial}{\partial r} \left(r \frac{\partial \phi}{\partial r} \right) + \frac{1}{r^2} \frac{\partial^2 \phi}{\partial \theta^2} + \frac{\alpha_r}{k_r} e^{-(r/\beta)^2} = 0 & \text{for the solid} \end{cases} \tag{16}$$

Even small linear perturbations must satisfy the equations above and an integer number of oscillations (first order approximation) will constitute the zero level-set. Assuming a starting contour equal to a circle of radius R the curvature will be equal to $\kappa = -\frac{1}{R}$ The time evolution of the level-set accordingly to the aforemention equations is represented in Fig. 3 where a very good agreement of the zero level-set with real data is reached. For further details about analytical integration of eq. 16 refer to (Hennessy (2010)).

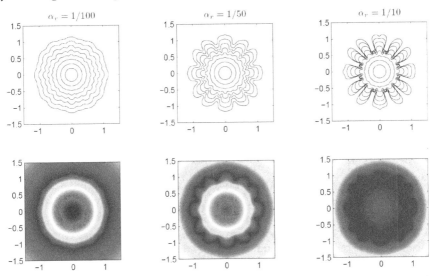

Fig. 3. on the top raw: different snapshots of the zero level set evolution varying the relative absorbition α_r; instead on the bottom line different level set values (different temperatures) are presented for the larger curve of the raw above

4. Motion planning for traveling robots in complex environments

Fluid dynamics and Level-set formulation can also be fruitfully adopted in Motion planning Algorithms. In particular the flexibility of the presented technique for time optimal motion planning with moving obstacles can be easily integrated with further constraints like different terrains traversability, path narrowing or fuel economy. Many classical techniques are present in literature for shortest path search with stationary obstacles, a good survey can be found in Latombe (1991). The space-time representation of obstacles moving in a 2D environment, assuming that one axis will be the time variable will be a solid structure. In Fig. 4 there is the representation of a circular obstacle moving in the x-y plane along x direction with an abrupt slow down.

As shown in Kimmel et al. (1998) the search for the time-optimal path within the aforementioned 3D time-space can be reduced to the search in a reduced set of 2D regions forming a Multivalued Distance Map (MDM). A Distance Map from a point **p** is a function that assigns to each point **q** in the considered domain a value corresponding to the "minimal length" (geodesic). The simplest approach uses a contour with a constant speed propagating from a small circle around **p** and the distance assigned to each crossed point **q** is proportional to the propagation time. If obstacles are moving the optimal path will be the minimal geodesic only when it does not cross any moving obstacles in the scene. For all the other cases a

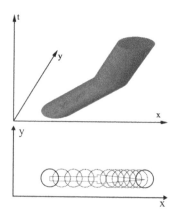

Fig. 4. a space-time representation for a circular obstacle moving in the x-y plane along the x direction that slow down its speed

more sophisticated approach is required: MDM will keep track of multiple crossing of the propagating contour when moving obstacles are present: in particular $MDM_p(\mathbf{q}, 1)$ will indicate the first time at which the point $\mathbf{q}$ is reached from $\mathbf{p}$ while $MDM_p(\mathbf{q}, 2)$ will indicate the second time at which, accordingly to the moving objects the propagating contour can reach $\mathbf{q}$, and so on. We will call $\mathcal{N}(\mathbf{q})$ the number of times that the propagating front will cross that point. Once the MDM is built for all points in the domain, the best path is obtained using a back-track procedure.

Consider now, given the source point $\mathbf{p}$, a continuous surface $\mathcal{S}$ in the space-time domain, centered in $\mathbf{p}$ indicating the wavefront contour as a function of time: $\mathcal{S} = \{(x, y, t(x, y))\}$, where a contour at a specific geodesic distance $\bar{d}$, $\alpha(\bar{d})$ is given by:

$$\alpha(\bar{d}) = \{\mathbf{q} | d_g(\mathbf{p}, \mathbf{q}) = \bar{d}\} \tag{17}$$

where $d_g(\mathbf{p}, \mathbf{q})$ is the minimal geodesic distance along the surface $\mathcal{S}$. Accordingly to Kimmel et al. (1995), due to the continuity and smoothness of $\mathcal{S}$ we can define a curvilinear abscissa u for $\alpha(u, \bar{d})$.

If we assume that the motion velocity is constant, V without loss of generality we can also assume that $V = 1$ then the geodesic distance $d = d_g(\mathbf{p}, \mathbf{q})$ coincides with time t and $\alpha(u, d) = \alpha(u, t)$ the equal distance contour evolution rule is:

$$\frac{\partial \alpha(u, t)}{\partial t} = \mathbf{N} \times \mathbf{T} \tag{18}$$

where $\mathbf{T}$ is the unit vector tangent to α while $\mathbf{N}$ is the unit vector normal to the surface $\mathcal{S}$.

To avoid point discontinuities the starting contour $\alpha(u, 0)$ is a small circle around the point $\mathbf{p}$ and the projection of equation 18 on the x-y plane will be the velocity vector orthogonal to the actual front, V_N:

$$V_N = \mathcal{P}(\mathbf{N} \times \mathbf{T}) \cdot \mathbf{n} \tag{19}$$

where $\mathcal{P}()$ is the projector on the x-y plane and $\mathbf{n} = (n_x, n_y)$ is the unit normal to the planar curve in the x-y plane.

The evolution of the planar contour will then be:

$$\frac{\partial P\left(\alpha\left(t\right)\right)}{\partial t} = -V_N \mathbf{n} \tag{20}$$

In Fig. 5 there is a graphical representation of the aforementioned quantities.

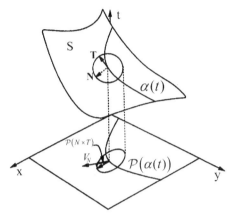

Fig. 5. The time evolution of variable α, and the definition of the moving front velocity V_N

Calling $p = \frac{\partial t(x,y)}{\partial x}$ and $q = \frac{\partial t(x,y)}{\partial y}$, accordingly to (Kimmel et al. (1995)) we can write:

$$\mathbf{N} = \frac{(-p, -q, 1)}{\sqrt{1 + p^2 + q^2}} \tag{21}$$

and:

$$V_N = \sqrt{\frac{(1+q^2)\, n_x^2 + (1+p^2)\, n_y^2 - 2pq n_x n_y}{1 + p^2 + q^2}} \tag{22}$$

Since the minimal geodesics are perpendicular to the equal distance contours on the surface, we can track the optimal trajectory starting from the destination point and, considering the equal distance contour α crossing it we have to move backward thanks to the orthogonality between the optimal path and the equal distance contour:

$$\eta = -\left(\mathbf{N} \times \mathbf{T}\right) \tag{23}$$

Using an arc length parametrization for the curve α we can write $r = \frac{\partial MDM_p}{\partial x}, l = \frac{\partial MDM_p}{\partial y}$ and using a normalization function

$$\psi\left(p,q,r,l\right) = \frac{p^2 + q^2 + 1}{\sqrt{l^2\left(1+p^2\right) + r^2\left(1+q^2\right) - 2pqrl}} \tag{24}$$

and the backtracking rule will then consist in tracking, from destination point $\mathbf{q}$ over the surface $\mathcal{S}$, the best path β so that:

$$\frac{\partial \beta}{\partial t} = \eta = \psi\left(p,q,r,l\right)\left[r\left(1+q^2\right) - pql, l\left(1+p^2\right) - pqr, pr + ql\right] \tag{25}$$

If there are moving obstacles along the path the MDM_p becomes multivalued and it could be possible that some points change their index along the optimal path, to account for this all distance values must be evaluated at a candidate point.

4.1 The Level-set approach

Moving obstacles introduce new boundary constraints and dealing with them could be very complex following the previously describer approach, a simpler way could be using the level-set approach where the projection of α on the x-y plane is the zero level set $\phi = 0$ and ϕ will be negative in the interior and positive in the exterior of the zero level set.

$$\phi \left(\mathcal{P} \left(\alpha \left(t \right) \right), t \right) = 0 \tag{26}$$

A graphical representation of this process can be seen in Fig. 6.
The next step requires the definition of the evolution law for ϕ, that, accordingly to the chain rule can be written as:

$$\nabla \phi \left(\mathcal{P} \left(\alpha \left(t \right) \right), t \right) \frac{\partial \mathcal{P} \left(\alpha \left(t \right) \right)}{\partial t} + \frac{\partial \phi \left(\mathcal{P} \left(\alpha \left(t \right) \right), t \right)}{\partial t} = 0 \tag{27}$$

Furthermore, accordingly to this level-set formulation, the in-plane normal vector can be written as $\mathbf{n} = \frac{\nabla \phi}{\|\nabla \phi\|}$. This relation together with the Lagrangian evolution equation:

$$\frac{\partial \phi \left(\mathcal{P} \left(\alpha \left(t \right) \right), t \right)}{\partial t} = -V_N \|\nabla \phi\| = \sqrt{\frac{\left(1 + q^2\right) \phi_x^2 + \left(1 + p^2\right) \phi_y^2 - 2pq\phi_x\phi_y}{1 + p^2 + q^2}} \tag{28}$$

This is the Euierial formulation for curve evolution.

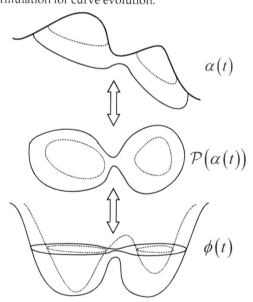

Fig. 6. Implicit representation of the projection of the α function through the level-set.

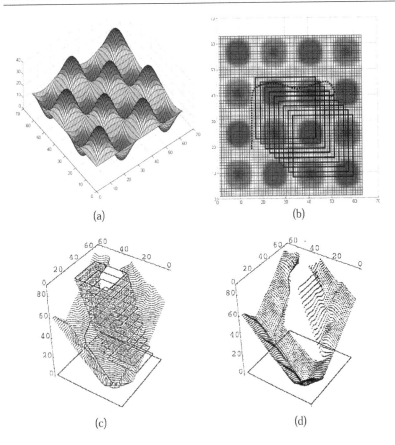

(a)   (b)

(c)   (d)

Fig. 7. (a) the terrain definition, (b) a squared object moves across the scene. (c) $\alpha(u,t)$ for different values of t representation (d)Back tracking from the final point toward the starting one along the estimated geodesic path

The steps for the best path definition can then be summarized as:

1. Evolve the PDE (eq.28) till the destination point is reached.

2. back-track along the optimal path till the origin.

In Fig. 7 there is the representation of level-set evolution, accordingly to moving objects, in an "egg-box" terrain where a square is moving across the scene.

Terrain traversability can be easily integrated in the presented approach since it is possible to define a map $F(x,y)$ describing the terrain traversability according to continuous values from 1 (optimal traversability) to 0 (impassable) and the optimal path can be obtained just changing eq. 18 into

$$\frac{\partial \mathcal{P}(\alpha(t))}{\partial t} = -F(x,y)\,V_N\mathbf{n} \tag{29}$$

Further velocity dependencies, e.g. from the vehicle weight reduction due to fuel combustion, can be accounted introducing a time dependance in V_N. More complex conditions can be easily added just acting on V_N, e.g. if the Robot has a specific shape the velocity can reflect the ability to walk through a narrow path or to turn at a specific point. In some cases the $F(x, y)$ correcting function is called viscosity for it similarity to fluid property that can prevent or slow down motion in particular circumstances.

5. Surface reconstruction from cloud of points

The temporal evolution of a volumetric implicit function can also be fruitfully applied to reconstruct 3D surfaces form large set of unorganized sample points. The evolving front can be thought as the surface that separates two different fluids obeying specific fluid dynamics laws. One remarkable feature of this approach is its ability to model complex topologies using a set of intuitive tools derived from fluid physics: Global and local surface descriptors are used allowing the parallelization of the algorithm on different object parts working with different resolutions and accuracies. The problem of building surfaces from unorganized sets of 3D points has recently gained a great deal of attention. In fact, in addition to being an interesting problem of topology extraction from geometric information, its applications are becoming more and more numerous. For example, the acquisition of large sets of 3D points is becoming easier and more affordable using, for example, 3D-scanners (*Registration and fusion of intensity and range data for 3d modelling of real world scenes*   (2003)). The presented approach is particularly important in applications where objects are better described by their external surface rather than by simple unorganized data (clouds of points, data slices, etc.). For example, in medical applications based on Tomography scans or NMRs it is often necessary to visualize some specific tissues such as the external surface of an organ starting from the acquired 3D points. This can be achieved by selecting the points that belong to a specific class (organ boundary, tissue, etc.) and then generating the surface from their interpolation. In most cases the definition of this surface is an ill-posed problem as there is no unique way to connect points of a dataset into a surface, therefore it is often necessary to introduce constraints for globally or locally controlling the surface behavior. As a matter of fact, the resulting surface often turns out to exhibit a complex topology due to noise in the acquired data or ambiguities in the case of non-convex objects Hoppe (1994). In order to overcome such problems, surface wrapping algorithms need to incorporate specific constraints on the quality of the data fitting (surface closeness to the acquired points), on the maximum surface curvature and roughness, on the number of resulting triangles, etc. The existing surface reconstruction methods can be classified into two broad categories: the former describes the surface as an implicit function while the latter describes it in an explicit form. Explicit (boundary) representations describe the surface in terms of point connections, and traditional approaches are based on Delaunay triangulation, Voronoi diagrams (Amenta et al. (1998)) or NURBS (Piegel & Tiller (1996)).

Here we will show how Level-set technique together with Volume of Fluid end Fluid Dynamics equations can be fruitfully applied to obtain a wrapping surface with intuitive parameters tuning. To evaluate quality of the final reconstruction an important parameter that can also be used as stopping condition for the surface evolution is an energy functional defined as:

$$E(\Gamma) = \left[\int_{\Gamma} d^p(\mathbf{x}) \, ds \right]^{1/p}, 1 \leqslant p \leqslant \infty \qquad (30)$$

where Γ represents the surface, $\mathbf{x}$ is the set of 3D points, d is the euclidean distance between the point $\mathbf{x}$ and the closest point on the surface Γ and $1 \leq p \leq \infty$ is the order of the functional distance: for $p = 2$ it gives a root mean square error while for $p = 1$ we just focus on the point with the maximum distance from Γ. looking for the surface that minimizes the functional (30) is similar to enveloping the data set with a membrane with certain elastic properties and have this membrane to evolve in time until it comes to rest. This time evolution paradigm can be described as:

$$\frac{\partial \Gamma}{\partial t} = - \left[\int_{\Gamma} d^p(\mathbf{x}) \, ds \right]^{\frac{1}{p}-1} d^{p-1}(\mathbf{x}) \left[\nabla d(\mathbf{x}) \cdot \mathbf{n} + \frac{1}{p} d(\mathbf{x}) \kappa \right] \mathbf{n} \tag{31}$$

Euler-Lagrange equation can the be used to find its minimum:

$$d^{p-1}(\mathbf{x}) \left[\nabla d(\mathbf{x}) \cdot \mathbf{n} + \frac{1}{p} d(\mathbf{x}) \kappa \right] \tag{32}$$

where $\mathbf{n}$ points accordingly to the evolving point direction, κ indicates the mean curvature, $\nabla d(\mathbf{x}) \cdot \mathbf{n}$ represents the surface attraction term and $d(\mathbf{x}) \kappa$ is the surface tension. An important consideration is that function $d(\mathbf{x})$ will automatically impose that the final surface is more flexible where the 3D points dataset is denser and more rigid in low density regions: this will allow smoother regions where we have less samples and accurate fitting in highly sampled regions. Fluids, with their properties like surface tension and viscosity represent an optimal modelization for the warping surface providing us with a flexible tool for this goal. Also in this case we present a Level-set approach, in particular we will adopt an implicit function ϕ over the spatial 3D domain so that $\phi(\mathbf{x}, t) = 0$ represent the warping surface $\Gamma(t)$ at time t, for the initialization we use $\phi(\mathbf{x}, t) \,|\, t = 0 = D$ where D is the signed distance of $\mathbf{x}$ from $\Gamma(t)$ (positive values indicate external points while negative ones represent internal points). One possible approach to make surface Γ to converge towards the cloud of points is, accordingly to Zhao et al. (2001), to use the fluids convection defining a velocity field $\mathbf{v}(\mathbf{x})$ accordingly to:

$$\frac{\partial \Gamma(t)}{\partial t} = \mathbf{v}(\Gamma(t)) \tag{33}$$

In particular we adopted, as a steering potential field, the distance function $d(\mathbf{x})$ so that the velocity field $\mathbf{v}$ is oriented towards the minima of the potential field: $\mathbf{v} = -\nabla d(\mathbf{x})$. So, accordingly to eq. 8, the level-set evolution in this case can be defined as:

$$\frac{\partial \phi}{\partial t} = \nabla d(\mathbf{x}) \cdot \nabla \phi \tag{34}$$

A major computational drawback comes from the fact that the ϕ function at each time iteration becomes more and more compressed presenting a growing slope across the zero level-set which could introduce numerical imprecisions. To solve this problem some solutions are present in literature, for example a level-set re-initialization in a region around the zero level: this is particularly useful for Fast Marching methods which, as described in section 3 for the Narrow Band approach Sethian (1999). Other approaches, instead, suggest a different implementation for the Hamilton Jacobi PDE like Gomes & Faugeras (1999). In particular in

Marcon et al. (2008) a quite different interpretation for the level-set function is presented that aims to conciliate Level-set upgrade, narrow band approach, discretization issues and Volume of Fluid (VoF) method: in particular, the VoF method (see Noh & Woodward (1976)) has been formulated in a variety of forms and re-introduced under different names such as the "cell method" and the "partial fractions method". The key idea behind this technique is to define a fixed computational grid and assign to each grid cell a value that describes the relative proportions of two materials contained in that cell. Our particular modeling metaphor is based on two immiscible fluids of opposite densities $\rho = 1$ and $\rho = -1$, which identify the surfaceŠs outside and inside, respectively. More specifically, cells completely filled with outer fluid take on the value +1 while cells that are only filled with inner fluid will take on the value -1. When filled with a mix of both fluids, the cell is assigned an intermediate value in the $]-1, 1[$ range, which measures the relative density of the mix (i.e. a value of zero indicates the same amount, half and half of the two fluids inside the cell). As a first step we define a bounding box that completely encloses the point cloud. We assume that this box is completely filled with inner fluid and that the whole space outside of the box is filled with outer fluid. We define the rules of evolution so that both fluids are set in motion by the presence of data points, which act as attractors. This fluid migration takes place in compliance with laws of conservation. According to such rules, the front starts propagating inward (towards the cloud of points) and stops only when the inner fluid is confined "inside" the cloud of points and the interface between the two fluids wraps the point-cloud. At each iteration step we will replace the value of points on the domain boundary $\mathcal{B}$ with 1 (this is equivalent to place outer fluid sources on the domain boundary), the cells with values greater than 1 will be clipped with value '1' while cells with values lower that -1 will be clipped to -1: this will be equivalent to remove excess fluids from cells without any need to introduce further fluid sources or drains. Two main contribution to fluid motion are considered: convective and diffusive: the first one is important for steering the zero level-set toward the cloud of points while the second has an important role in smoothing evolving front. Accordingly to the Gauss theorem when no sources or drains are present the fluid density will follow eq. (35):

$$\frac{\partial \rho}{\partial t} = \nabla \cdot (\mathbf{v}\rho) + \nabla \cdot (\xi \nabla \rho) \qquad (35)$$

where ρ indicates the fluid density, $\mathbf{v}\rho$ is the convective term indicating the amount of fluid per volume unit that is transported by the velocity field while $\xi \nabla \rho$ represents the diffusive part where ξ is the diffusive term (further details in (Gueyffier et al. (1999))). In this approach a point will be assigned a velocity that is proportional to the vector $\mathbf{v}$ that joins that point to the closest data point. This way data points act as attractors for both internal and external fluids and, therefore, for the evolving front: each sample point represents a central vector field. Even if this attraction field has no physical equivalent, it is physically consistent thanks to its irrotational characteristic (see Marcon et al. (2008) for further details). This fact guarantees algorithmic convergence. In Figures 8 and 9 it is possible to see the direction and intensity of the steering field toward the cloud of points and different level-set curves; in particular in the magnification of zones A and B is possible to see how the steering field automatically stops external liquid from entering the object (all the internal arrows point outside) while, in the convex part, the steering field force the outer fluid to enter the convex region.

In Fig. 10 we represent the final state for the wrapping of a two dimensional cloud of points.

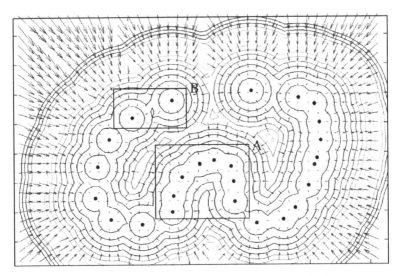

Fig. 8. a 2D non-convex example, arrows indicate the steering velocity field toward the cloud of points and different level-sets are also represented. Zones A and B are zoomed in Fig.9

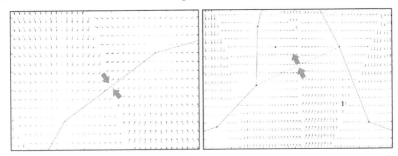

Fig. 9. zoom of zone A and B of Fig. 8 in particular different behavior of the steering field prevents external fluid from entering the convex part in B while favor its penetration in A (nonconvex region)

Further aspects from fluid dynamics can be fruitfully applied, e.g. the diffusive parameter ζ in eq. 35 is a delicate parameter as it is strictly connected to fluid viscosity. High values of ζ result in a smooth but often inaccurate surface while low values result in a surface that strictly honors the data but is sensitive to acquisition noise. As a matter of fact for laser scanner acquisitions the portions of the object with high samples density (usually due to multiple acquisitions from different viewpoints) requires high accuracy and/or present complex topologies, while smoother or less critical areas usually exhibit a lower sample density. In order to account for that, we can establish an inverse dependency between the diffusive constant ζ and the local point density:

$$\zeta = \frac{\alpha}{1 + \rho_p} \qquad (36)$$

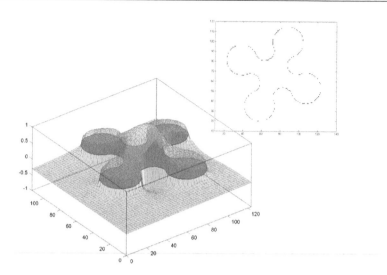

Fig. 10. the wrapping of a set of 2D points (top right), the zero level-set at the end of the evolution represent the joining line, inner points have positive ψ while external ones have negative values

where α is a parameter that indicates the estimated noise in the point location (the higher the value the smoother the surface). The density ρ_p at a given point is defined as the number of samples that fall into a cube of volume V/N around that point, where V is the volume of the point cloud (approximated by the volume of the smallest parallelepiped that contains the point cloud) and N is the total number of samples. This means that if all the samples would be uniformly distributed in the considered volume (like in a cubic crystal) in each cubic cell of volume V/N there would be just one sample. In our case, since we have to define the ρ_p variable for each voxel, we define it as the sum of all the points falling inside a cube of volume V/N centered at the voxel center. In surface regions with a low point density the diffusive constant ζ corresponds to a smooth surface, while in high-density zones ζ becomes very low, to guarantee that the surface will strictly honor the sample points.

The level-set evolution equation is then the following:

$$\begin{cases} \frac{\partial \phi(\mathbf{x},t)}{\partial t} = \nabla \cdot (\mathbf{v}\phi(\mathbf{x},t)) + \zeta \nabla^2 \phi(\mathbf{x},t), \, t \geqslant 0 \\ \phi(\mathbf{x},t) = +1, & t \geqslant 0, \mathbf{x} \in \mathcal{B} \\ \phi(\mathbf{x},0) = -1 & \mathbf{x} \notin \mathcal{B} \end{cases} \qquad (37)$$

where the second raw indicates that the boundary $\mathcal{B}$ of the considered domain is always kept equal to +1 assuming that those points are sources of external fluid. The third raw indicates that at the starting point all the points inside the domain are filled with the internal fluid. In Fig. 11 the level-set evolves toward the cloud of points which is a laser scanner acquisition of a plaster wolf.

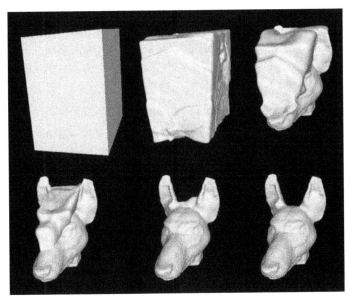

Fig. 11. Evolution of the zero level-set toward a 3D cloud of points acquired by a laser scanner. The level-set starting from the boundary of the considered volume (top left) evolves towards the final cloud of points.

5.1 Turbolence and convex regions

In some particular cases, where high curvature is present in nonconvex regions with a small entrance, external fluid could be inhibited from entering (e.g. consider the cloud of points of Fig. 12).

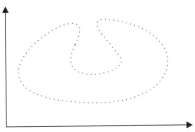

Fig. 12. a set of points presenting a high curvature region within a concavity

For cases like these a further help could come from vorticity. In particular, since our steering field is based on the distance from the closest point it will be a conservative irrotational vector field and when the outer fluid will approach the concavity mouth it will be pushed to its borders without the opportunity to get inside (we underline that this condition is different from the one in Fig. 8 where the concavity of the region A does not present high curvature and external fluid is correctly steered inside the region). A possible solution consists in introducing a rotational component in the steering field, this could be obtained adding periodically a small random displacement for each point around its original position: it will result in a vorticity contribution: $\zeta = \nabla \times \mathbf{v}(\mathbf{x})$ the resulting effect is that inner fluid in convex regions is pushed inside and outside with the same probability while in nonconvex regions inner fluid that went inside stays there and progressively fills the region. Particular care must be placed in introducing the amount of random points displacement since convergence of the algorithm is no more guaranteed by the conservative vector field.

6. Image processing and viscous noise removal

Fluid dynamics applied together with level-set method can also provide a powerful tool for noise removal in complex images (i.e. when noise is not simply removable using intensity thresholds of histogram analysis), a Total Variation denoising approach is presented in (Osher & Fedkiw (2002)) where the regularization algorithm is based on this PDE:

$$Lu = \lambda Ru, \tag{38}$$

where u is the image represented as a 2D function $(u(x,y))$ that indicates intensity value for each point; R is the regularization operator and L is the time-space operator. A very common energy functional $E(u)$ whose minimization provide the denoised image is the Mumford-Shah multiscale segmentation model:

$$\min_u E(u) = \frac{1}{2} \int_\Omega (u - u_0)^2 dx + \mu \int_{\Omega \backslash \Gamma} |\nabla u|^2 dx + v \mathcal{H}^2(\Gamma) \tag{39}$$

where u_0 is the given noisy image, Ω is the image function domain and μ is a weighting coefficient for the average smoothness of regions divided by the contour Γ. v is the weight coefficient for the total contour length expressed by the Haussdorf measure $\mathcal{H}$.

The minimization of $E(u)$ will provide an piecewise-smooth approximation of the initial u_0 image, Γ has the role of approximating the edges in the image u_0 and u will be smooth only in $\Omega \backslash \Gamma$; Accordingly to the level-set formulation the implicit function will be ϕ and in particular $\phi(\mathbf{x}) = 0$ represents the contour Γ separating region of positive values of ϕ from the region of negative values. In particular $u^+(\mathbf{x})$ will be the intensity function representing values of the region where $\phi(\mathbf{x}) > 0$ while $u^-(\mathbf{x})$ will represent the intensity of points in the $\phi(\mathbf{x}) < 0$ region. Eq. 39 then become:

$$\min_u E(u^+, u^-, \phi) = \int_\Omega (u^+ - u_0)^2 H(\phi) dx + \int_\Omega (u^- - u_0)^2 (1 - H(\phi)) dx +$$
$$+ \mu \int_\Omega |\nabla u^+|^2 H(\phi) dx + \mu \int_\Omega |\nabla u^-|^2 (1 - H(\phi)) dx + v \int_\Omega |\nabla H(\phi)| \tag{40}$$

Where $H()$ represents the Heaviside step function and the last term, $\int_\Omega |\nabla H(\phi)|$ is the integral in the sense of geometric measure, i.e. for a 2D case, is the length of contour Γ.

The minimization can be done iteratively on u^+, u^- and on ϕ, in particular, minimizing with respect to u^+ and u^-, i.e. deriving eq. 40, will give:

$$u^+ - u_0 = \mu \nabla^2 u^+ \tag{41}$$

and

$$u^- - u_0 = \mu \nabla^2 u^- \tag{42}$$

which correspond to the diffusive part inside the borders defined by Γ, since $\frac{\partial u^{+,-}}{\partial n} = 0$ (i.e. no diffusion will take place across the boundary keeping edges sharp). While, deriving with respect to ϕ will give:

$$\frac{\partial \phi}{\partial t} = \delta_\epsilon (\phi) \left[- \left| u^+ - u_0 \right|^2 - \mu \left| \nabla u^+ \right|^2 + \left| u^- - u_0 \right|^2 + \mu \left| \nabla u^- \right|^2 + \nu \nabla \left(\frac{\nabla \phi}{|\nabla \phi|} \right) \right] \tag{43}$$

Accordingly to a steepest descendent approach. The δ_ϵ indicates that the delta of Dirac, in a discrete environment, must be calculated in a narrow band and, to prevent discontinuities u^+ has to be extended, in the narrow band, for $\phi < 0$ values and also u^- has to be extended in the narrow band for $\phi > 0$. A possible approach to these extension could be the "Ghost fluid Method" (Fedkiw et al. (1999)) where the interface is treated as a moving boundary. Solving the governing equations (Terashima & Tryggvason (2009)) with the extension of the discontinuous variables across the fluids interface, (for real CFD applications this is usually entropy), will reduce smearing of discontinuities providing smooth convergence to the final image without oscillations. In Fig. 13 is represented how the contour between the two image parts is steered accordingly to the Ghost Fluid Method in the narrow band while in Fig. 14 it is possible to see the evolution of the contour across the noisy image: diffusive terms smooth uniform regions reducing noise while the contour evolution smoothly warp the objects.

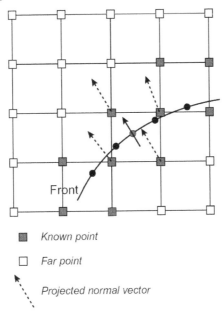

Fig. 13. update procedure in the narrow band points

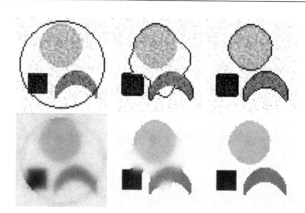

Fig. 14. zero level set evolution accordingly to the Mumford-Shah energy minimization for object segmentation and image denoising

7. Conclusion

In this chapter we have shown some significant implementations of Computational Fluid Dynamics equations in completely different contexts ranging from image denoising to optimal path research for moving robots. In recent literature it is also possible to find different applications to many other research fields: The main reason for this large diffusion and reuse of CFD equations can be mainly addressed to:

- the actual deep insight and knowledge both into the theoretical and computational aspect of fluids.

- The accuracy and stability of modelling equations that can grant convergence in a wide variety of applications.

- Versatility and adaptability of different models from multiphase, to viscous or compressible fluids that provide the user with many intuitive and easy-to-tune parameters allowing an easy adaptation to different contexts.

- The close relation to physical quantities that in many cases can automatically provide the minimum energy solution.

All these aspects concurred in making CFD a science that goes well beyond fluids analysis providing a self consistent, well known, spread of tools to find fruitful and unlimited applications in engineering, computer science and other scientific research fields.

8. References

Adalsteinsson, D. & Sethain, J. A. (1995). A fast level set method for propagating interfaces, *Journal of Computational Physics* 118(2): 269–277.

Amenta, N., Bern, M. & Eppstein, D. (1998). The crust and the β-skeleton: combinatorial curve reconstruction, *Graph. Models Image Process* 60(2): 125–135.

Biben, T., Klaus, K. & Chaouqi, M. (2005). Phase-field approach to three-dimensional vesicle dynamics, *Phys. Rev. E* 72(4).

Bridson, R. (2003). *Computational aspects of dynamic surfaces*, PhD thesis, University of British Columbia.

Davis, S. H. (2001). *Theory of solidification*, Cambridge University Press.

Dobrushin, R. L., Kotecký, R. & Shlosman, S. (1992). *Wulff Construction: A Global Shape from Local Interaction*, American Mathematical Society.

Fedkiw, R. P., Aslam, T., Merriman, B. & Osher., S. (1999). A nonoscillatory eulerian approach to interfaces in multimaterial flows (the ghost fluid method), *Journal of Computational Physics* 152(2): 457–492.

Gomes, J. & Faugeras, O. (1999). Reconciling distance functions and level sets, tech. rep. 3666, *Technical report*, INRIA Sophia Antipolis.

Gueyffier, D., Li, J., Nadim, A., Scardovelli, R. & Zaleski, S. (1999). Volume-of-fluid interface tracking with smoothed surface stress methods for three-dimensional flows, *J. Comput. Phys.* 152: 423–456.

Hennessy, M. G. (2010). *Liquid Snowflake Formation in Superheated Ice*, PhD thesis, University of Oxford.

Hills, R. N. & Roberts, P. H. (1993). A note on the kinetic conditions at a supercooled interface, *Int. Comm. Heat Mass Transfer* 20: 407–416.

Hobbs, P. V. (2010). *Ice Physics*, Oxford University Press.

Hoppe, H. (1994). *Surface reconstruction from unorganized points*, PhD thesis, University of Washington.

Kimmel, R., Amir, A. & Bruckstein, A. M. (1995). Finding shortest paths on surfaces using level sets propagation, *IEEE Transactions on Pattern Analysis and Machine Intelligence* 17(6): 635–640.

Kimmel, R., Kyriati, N. & Bruckstein, A. M. (1998). Multivauled distance maps fro motion planning on surfaces with moving obstacles, *IEEE Trans. on Robotics and automation* 14(3): 427–436.

Latombe, J. C. (1991). *Robot Motion Planning*, Kluwer.

Libbrecht, K. G. (2005). The physics of snow crystals, *Rep. Prog. Phys. 68*, Norman Bridge Laboratory of Physics, California Institute of Technology.

Losasso, F., F. Gibou, F. & Fedkiw, R. (2004). Simulating water and smoke with an octree data structure, *ACM Transactions on Graphics* 23(3): 457–462.

Marcon, M., Piccarreta, L., Sarti, A. & Tubaro, S. (2008). Fast pde approach to surface reconstruction from large cloud of points, *Computer Vision and Image Understanding* 112: 274–285.

Markstein, G. H. (1951). Experimental and theoretical studies of flame front stability, *J. Aero. Sci.* 18: 199.

Noh, W. & Woodward, P. (1976). A simple line interface calculation, *in* Springer-Verlag (ed.), *Proceedings of 5th International Conference on Fluid Dynamics*.

Osher, S. & Fedkiw, R. (2002). *Level Set Methods and Dynamic Implicit Surfaces*, Springer-Verlag.

Piegel, L. & Tiller, W. (1996). *The NURBS Book*, second ed. edn, Springer-Verlag, Berlin.

Pimpinelli, A. & Villain, J. (1999). *Physics of Crystal Growth*, Cambridge University Press.

Registration and fusion of intensity and range data for 3d modelling of real world scenes (2003). Vol. Fourth International Conference on 3-D Digital Imaging and Modeling, 3DIM Proceedings.

Sethian, J. A. (1989). A review of recent numerical algorithms for hypersurfaces moving with curvature-dependent speed, *Journal Differential Geometry* 31: 131–161.

Sethian, J. A. (1999). *Level Set Methods and Fast Marching Methods: Evolving Interfaces in Computational Geometry, Fluid Mechanics, Computer Vision, and Materials Science,* Cambridge University Press.

Shastri, S. S. & Allen, R. M. (1998). Method of lines and enthalpy method for solving moving boundary problems, *International Communications in Heat and Mass Transfer* 5(4): 531–540.

Terashima, H. & Tryggvason, G. (2009). A front-tracking/ghost-fluid method for fluid interfaces in compressible flows, *Journal of Computational Physics* 228: 4012–4037.

Zhao, H., Osher, S. & Fedkiw, R. (2001). Fast surface reconstruction using the level set method, *Proceedings of IEEE Workshop on Variational and Level Set Methods in Computer Vision (VLSM 2001).*

9

Fluid Dynamics in Space Sciences

H. Pérez-de-Tejada
Institute of Geophysics, UNAM
Mexico

1. Introduction

Much of what it is understood in nature and that it is also inherent to our common activities can be appropriately interpreted as representing evidence of a collective response that substantiates the basis of fluid dynamics. Such is the case for the behavior for a group of particles or individuals that interact with each other as they follow common trajectories. Under most circumstances their interaction takes place across distances (mean free path) that are much smaller than the size of the region where they move and, as a whole, they exhibit properties (density, speed, temperature) that represent average local values of the conglomerate. Besides the application of this view to standard problems in physics and engineering it has been intuitively suggested that it could also be relevant to describe the motion of the solar wind which at large distances from the sun travels with its particles barely executing any collisions among them. Even though the solar wind is, in fact, a (collisionless) ionized gas it maintains a collective response as it expands through interplanetary space and interacts with the planets of the solar system. The text below describes the manner in which the fluid dynamic interpretation of the solar wind was initiated and how it has expanded to examine its interaction with the planets of the solar system.

2. The solar wind as a continuous expanding gas

The over one million degree temperature of the solar corona is significantly larger than the nearly six thousand degrees of the solar surface and thus provides the energy source that ultimately leads to its strong outward expansion. In fact, the large amounts of energy that are delivered to the solar corona from the solar interior can be mechanically used for producing the solar wind. The prediction and theoretical description of this phenomenon was provided by E. N. Parker [1] at the University of Chicago who over 50 years ago devised a fluid dynamic interpretation of the manner in which the ionized coronal gas (plasma) is forced to expand outward reaching speeds of the order of 300-400 km/s. A remarkable aspect of this concept is that the solar wind is predicted to be an outflowing continuum gas that expands away from the sun as its density decreases to values that by the earth´s orbit are very small (~ 10 cm^{-3}). As a result, collisions among its particles (mostly protons and electrons) become extremely rare and the solar wind rapidly becomes a collisionless plasma with an effective mean free path for the collisions of its particles that is comparable to one astronomical unit (this is the distance between the earth and the sun). Despite this constraint observations made with various spacecraft in the interplanetary medium have confirmed the existence of the solar wind and its overall collective response when it interacts with the planets of the solar system. Much research has been conducted to investigate the processes that allow the solar wind to explain its collective behavior as it moves through space, interacts with the magnetic fields and the ionospheres of planets, and reaches its outer boundary (the heliopause) located at nearly 80 astronomical units.

A fundamental property of the solar wind is that it is rapidly accelerated to reach supersonic speeds and that the process that produces it was incorporated by applying conditions similar to those employed in the fluid dynamic de-Laval nozzle theory. The latter problem serves in rocket engine turbines and is designed to produce supersonic speeds in a streaming flow that is subject to a pressure gradient along its direction of motion. When a subsonic flow is compressed by decreasing the cross-sectional area of the region where it moves its speed increases but remains subsonic. In order to make the flow supersonic it is necessary to introduce a nozzle (converging/diverging) geometry so that the gas first becomes compressed and later expands upon streaming through a "throat" or region of minimum cross-sectional area as it is depicted in Figure 1. The reason here is that when a flow moves into the region of smaller cross-sectional area the imposed pressure gradient can increase the flow speed only up to the sound speed, but a further increase requires an

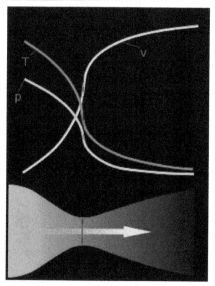

Fig. 1. Scheme of flow through a de-Laval nozzle. The speed profile (blue curve) shows how the flow reaches the sound speed at the throat and becomes supersonic afterwards. The temperature and pressure profiles (labeled T and P) indicate how the gas cools off and expands when it becomes supersonic [2].

expansion, that is, that the flow proceeds in a diverging geometry. These considerations are better illustrated by the de-Laval equation in hydrodynamics[2]:

$$ds/s = (v^2/c^2 - 1)\, dv/v \tag{1}$$

where v is the speed of the flow, s is the cross-section of a duct or region where the flow moves and c is the speed of sound (dv and ds are changes of those variables along the direction of motion). This equation is derived from the continuity and the momentum equations of the flow and states the correlation that there exists between a change in the speed of the flow and that of its cross-sectional area. Most notable is that when the duct is converging (negative ds < 0 values) and the speed of the flow increases (positive dv > 0 values) then v must be less than c so that the value of the parenthesis is negative. Such conditions correspond to subsonic flows where the Mach number M = v/c is smaller than one. When v = c, that is if M = 1, we now have ds = 0 and the duct should stop converging. Finally, if v is to exceed c (supersonic flows) then ds must increase (ds > 0) for positive changes in the value of the flow speed (dv > 0), thus implying that the duct should now be diverging. Such flow conditions of the de-Laval nozzle require, in addition, a large pressure difference along the direction of motion. If that difference is not large, implying that the speed of the flow is slow and does not reach the sound speed at the narrowest part of the duct, then it will only speed up the flow as a Venturi tube achieving only subsonic speeds.

The coordinated dependence between changes in the speed of the flow and those in the cross-section of the duct that was derived from equation 1 for the converging/diverging geometry of Figure 1 is a remarkable property that characterizes subsonic and supersonic flows. Its origin lies on the simultaneous consideration of pressure forces that balance the momentum convective term in the momentum equation of the flow from which Equation 1 was derived. On the other hand, even though the presence of the gravitational force and the outward radial expansion of the solar wind in the solar corona leads to flow conditions that are entirely different from those of the two-dimensional flow that streams in the duct of Figure 1, it is significant that a relationship similar to that in Equation 1 can also be derived by employing the continuity and the momentum equations that apply to the solar wind. Following Parker it is possible to replace equation 1 by [3]:

$$[2 - MG/(r\,c^2)]\,dr/r = (v^2/c^2 - 1)\,dv/v \qquad (2)$$

where M is the solar mass and G the gravitational constant in this relation which holds along the distance r away from the sun. This equation was derived by considering that the gravitational force constricts or chokes the solar wind in a manner similar to the converging section of the de-Laval nozzle in Equation 1; that is, the gravitational force replaces the geometric throat that is not present in the region where the solar wind expands. All in all it is very significant that in the derivation of Equation 2 the gravitational force plays a role similar to that of the geometric throat in Equation 1. The similitude between equations 1 and 2 serves to substantiate the fluid dynamic interpretation of the motion of the solar wind. For example, if the expansion velocity of the solar wind increases outward (dv > 0) then for subsonic speeds (v < c) we need $MG/(c^2r) > 2$ so that both sides of the equation become negative, and for supersonic speeds (v > c) we need $MG/(c^2r) < 2$ in order to make them positive. These conditions imply that there will be subsonic speeds at small r values and supersonic speeds at large r values. The speed of the solar wind will reach the speed of sound at the critical distance $r_c = MG/2c^2$ implying that the solar wind will gradually become supersonic as it moves away from the sun. Measurements conducted with various spacecraft have confirmed this result and, in fact, its observed speed is supersonic being about eight times the speed of sound. It should be noted here that the solar wind will become supersonic at a distance between three and four solar radii, or about two million kilometers above the sun surface. In addition, since the critical distance decreases with increasing the temperature T of the coronal gas, that is by increasing the speed of sound $c = \sqrt{(\gamma KT/m)}$ where γ is the specific heats of the gas, K the Boltzmann constant and m the mass of the gas particles, then it may occur that the critical distance becomes located below one solar radius if the gas temperature is very high (larger than four million degrees). In that case the solar wind will not reach the sound speed but will remain subsonic as it expands. Under standard conditions the solar wind becomes supersonic and flows as a continuum gas whose motion is also related to the magnetic field fluxes that are brought up from the sun. As it will be shown below the effects of motion of the solar wind particles in the presence of a magnetic field can be better described by considering the interaction of the solar wind with the earth´s magnetic field.

3. The gas dynamic analogue of the solar wind around the earth´s magnetosphere

An important contribution to the continuum flow interpretation of the solar wind became available from studies directed to describe its interaction with the earth´s magnetic field. That field is, in fact, the effective obstacle that the solar wind encounters as it approaches the earth since its ionized populations (protons and electrons) can be directly influenced by the planetary magnetization. The effect of their interaction is that as a whole the solar wind compresses and deforms the distribution of the earth´s magnetic field which becomes bounded in the dayside and is deviated in the nightside forming a magnetic tail that extends far downstream from the earth as it is illustrated in Figure 2. As a result the earth´s magnetic

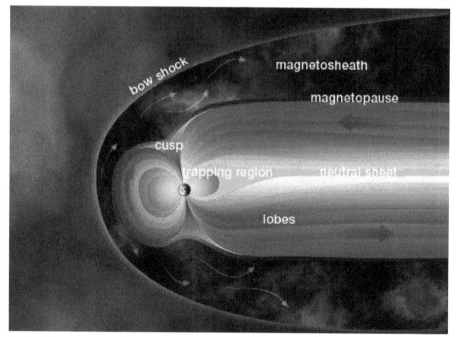

Fig. 2. A schematic view of the earth's magnetosphere formed by the supersonic solar wind that compresses the magnetic field and streams around it as an obstacle after crossing a bow shock front [3, 4].

field becomes present only within a large confined region in the form of a cavity (called the magnetosphere) whose outer boundary may reach seven earth radii in the direction of the sun. That boundary (the magnetopause) results from a local pressure balance between the solar wind pressure and the compressed pressure of the earth´s magnetic field, and implies a collective response despite the fact that there are no particle collisions in the solar wind population. The mechanism that replaces collisions to allow the transport of information upon encountering the earth´s magnetic field is the gyration that the protons and electrons of the solar wind execute when they enter the region occupied by the magnetic field. The process is related to the magnetic (Lorentz) force produced by the magnetic field and the velocity of the particles, and that drives them in a circular (Larmor) motion in a plane transverse to the magnetic field direction. Implied by this motion there is an electric current along the boundary of the magnetosphere that is generated by the different trajectory of the protons and the electrons of the solar wind as they interact with the earth´s magnetic field. Their accelerated motion around the magnetic field lines is important because it produces an induced magnetic field whose presence influences the motion of other particles. Thus, despite the fact the solar wind particles are not interacting through collisions with each other there is an ample communication among them which is produced by their motion around the earth´s magnetic field lines and that, in the end, justifies its collective response to the presence of that field.

This general idea to view the encounter of the solar wind with earth´s magnetic field led J. Spreiter [4] and M. Dryer [5] to apply a fluid dynamic model available from studies of the interaction of a streaming object in the earth´s atmosphere. The purpose of their studies was to describe the manner in which the solar wind encounters the earth´s magnetic field and then streams around the magnetospheric cavity. The gas dynamic description of the fluid properties of the solar wind as it reaches the earth´s magnetic field and then streams around the magnetosphere is dominated by various features. Most important is a bow shock produced by the supersonic speed of the solar wind and that is located upstream from the magnetosphere in agreement with the common response of a supersonic flow that encounters an obstacle. As it occurs in standard fluid dynamic problems the solar wind decelerates across the bow shock where it also becomes compressed and thermalized reaching values for these properties that are in agreement with those expected from the Rankine-Hugoniot equations of fluid dynamics. As the solar wind streams behind the bow shock and moves around the magnetosphere it gradually recovers its freestream conditions by expanding, cooling and increasing its flow speed as it has been observed with various spacecraft that have probed those regions. As a whole the changes that the solar wind experiences as it interacts with the earth´s magnetic field fit adequately with those expected in fluid dynamics.

An evolution of the solar wind similar to that observed around the earth´s magnetosphere also occurs when it encounters the magnetic field of other planets. Such is the case for Jupiter, Saturn Uranus and Neptune where their strong intrinsic magnetization leads to large scale magnetospheric cavities and that may reach up to 100 Jovian radii along the sun direction upfront from Jupiter. While comparable large values are also measured for the other planets conditions at Mercury are different where its magnetospheric cavity is very small and only reaches a few thousand kilometers above the surface in the direction of the sun. The solar wind that streams around the magnetic cavity of all these planets bears a similar response in the interaction process.

4. The fluid dynamic interaction of the solar wind with the upper ionosphere of Venus and Mars

4.1 Transport of solar wind momentum to the Venus ionosphere

Unlike magnetized planets Venus and Mars present obstacles to the solar wind that lead to phenomena that are not related to its interaction with a planetary magnetic field. In particular, in the absence of an appreciable intrinsic planetary magnetization the solar wind does not stream around a magnetic cavity but reaches the upper atmosphere of the planets. The flow conditions are more adequately described as an interaction between two separate bodies of plasma namely, the solar wind and the upper ionosphere. Consequently, the dynamics of the particles that are involved in their interaction is entirely different from that occurring at the boundary of a planetary magnetosphere, and to a large extent it is far more complicated. For example, the access that the solar wind acquires when interacting with particles in a planetary upper ionosphere produces phenomena that do not occur when the solar wind interacts with a planetary magnetic field and suggest the existence of turbulent flow conditions. Information obtained from measurements conducted by spacecraft that have probed Venus and Mars indicate that the interaction between both ion populations seems to be more strongly influenced by turbulent fluid processes rather than through a plasma dynamics similar to that occurring at the boundary of the earth´s magnetosphere. There is, in fact, evidence obtained from different sources that suggest the existence of phenomena that strongly resemble those occurring in fluid dynamic problems. A collected view of the various features that have been detected in the Venus plasma environment is presented in Figure 3 from observations conducted with the Pioneer Venus spacecraft (PVO) [6]. They include, in

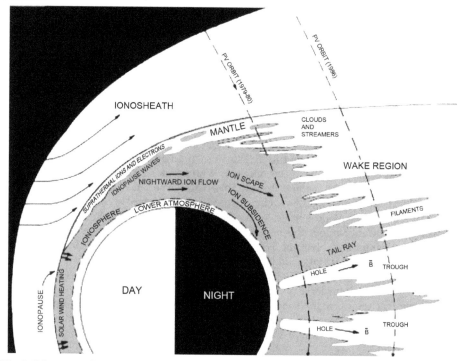

Fig. 3. Schematic diagram of plasma features (bow shock, ionopause, nightward ion flow, tail rays, ionospheric holes) that were identified near Venus with instruments in the Pioneer Venus Orbiter spacecraft (the altitude scale has been enlarged to better describe the geometry of the plasma features)[6].

addition to a bow shock, plasma clouds or regions of separated ionospheric plasma that imply, in fact, crossings through structures or filaments that extend downstream from Venus. At the same time there are regions within the nightside ionosphere where there is a large deficiency in the plasma density (ionospheric holes), and equally important is a generalized nightward directed flow of the dayside ionospheric plasma across the terminator.

Studies of the origin of the latter phenomenon (trans-terminator ionospheric flow) have been directed to examine measurements conducted in the region where it has been observed. From the early Mariner 5 spacecraft mission that probed the Venus environment it was learned that the solar wind that flows around the ionosphere exhibits a significant loss of its momentum; that is, its velocity and density show a strong deficit with respect to the freestream values [7]. That variation is described in the velocity and in the density profiles of the Mariner 5 spacecraft that are shown in the upper panel of Figure 4, and imply that a large fraction of the oncoming energy (shaded regions along the wake) has been removed

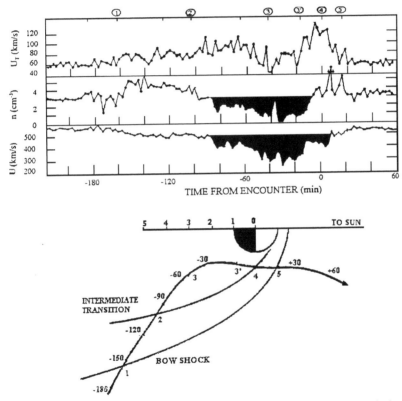

Fig. 4. Thermal speed, density, and bulk speed of the solar wind measured with the Mariner 5 spacecraft (its trajectory projected in cylindrical coordinates is shown in the lower panel). The labels 1 through 5 along the trajectory and at the top of the upper panel mark important events in the plasma properties (bow shock, intermediate plasma transition) [7].

from the solar wind. Related information was obtained later from measurements conducted with the Pioneer Venus Orbiter which revealed the ionospheric flow across the terminator that is indicated in the upper panel of Figure 5 and that reaches 3-4 km/s speeds as it is shown in the speed profile in the lower panel of that figure [8,9]. The momentum implied by

ION DYNAMICS IN THE VENUS IONOSPHERE

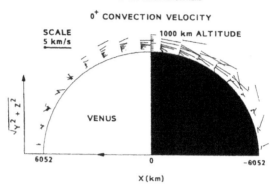

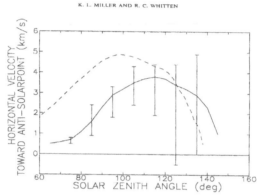

Fig. 5. (upper panel) Average velocity vectors measured in the Venus upper ionosphere with the PVO spacecraft and projected in cylindrical coordinates (the solar wind arrives from the left) [8]. (lower panel). Speed values of the ionospheric particles that move across the terminator at 400 km altitude[9].

that ionospheric flow is related to the deficit of momentum of the solar wind measured outside the flanks of the ionosphere that is shown in Figure 4. In fact, its kinetic energy density (momentum flux) is comparable to the quantity that was removed from the solar wind; that is, it implies an efficient transport of solar wind momentum to the Venus upper ionosphere [10]. Calculations of a momentum conservation equation applied to this problem led to such result which is unrelated to pressure gradient forces across the terminator that have also been suggested to account for the Venus ionospheric flow [9]. The (supersonic) speed of the ionospheric flow and its asymmetry in latitude which is indicated by differences in the extent of the region where the ionospheric plasma is displaced downstream from Venus argues more favorably in terms of momentum transport which mostly occurs by the magnetic polar regions of the Venus ionosphere [11]. It should be noted in this regard that the "magnetic polar regions" are related to the position where the solar magnetic field fluxes that are convected by the solar wind, and that first pile up around the dayside ionosphere to form a magnetic barrier [12], slip over the planet to take the shape of a hairpin at those regions and then enter the wake[13]. Thus, the outcome of the transport of solar wind momentum to the Venus upper ionosphere stresses the value of a fluid dynamic view to interpret the interaction between both bodies of plasma despite the fact that it is still necessary to define the mechanisms that through turbulent flow conditions support that behavior.

4.2 Plasma transitions outside the Venus ionosphere

Together with the transport of momentum into the Venus ionosphere the presence of boundaries in the plasma data shown in Figure 4 is indicative of fluid dynamic processes that are operative throughout the region of interaction with the solar wind. As the Mariner 5 spacecraft moved near Venus in its flyby transit from the wake to the dayside (its trajectory projected in cylindrical coordinates is shown in the lower panel of Figure 4) there is clear evidence of two crossings of a bow shock at events labeled 1 and 5. Across the bow shock the speed of the solar wind is smaller in the downstream side (bottom profile in the upper panel), and there is also evidence of heating (larger values in the thermal speed profile labeled U_T) and higher density values (all these variations are similar to those encountered across the bow shock present upstream from the earth's magnetosphere in Figure 2). Equally informative is the observation of a different plasma transition at events labeled 2 and 4 which indicate changes that are not related to a bow shock crossing. In particular, while the solar wind speed further decreases in the downstream side of that transition the density does not increase but exhibits lower values (after event labeled 2 and before event labeled 4). Thus, rather than a bow shock crossing where the solar wind becomes compressed the changes at that plasma transition indicate the presence of a region where the plasma becomes expanded. That transition is not the upper boundary of the ionosphere (the ionopause) which was not crossed by the Mariner 5 spacecraft but a boundary that is innovative since it has not been detected downstream from the bow shock of planetary magnetospheres.

Arguments leading to the presence of the latter transition do not follow from standard views of the acceleration of planetary ions that are generated through the ionization of neutral particles of the Venus atmosphere and that are immersed in the solar wind. In those views the neutral particles that form the planet's exosphere become ionized by the ultraviolet and the x-ray solar radiation and as a result should be rapidly accelerated by an electric field that arises from the relative speed between them and the solar wind [14]. The process, named mass loading, is expected to account for the population of planetary ions that is carried by the solar wind and whose contribution should gradually increase with the larger density of exospheric neutral particles that the solar wind encounters when it approaches the Venus atmosphere. While there is clear evidence for the existence of such process there is no indication that it should produce a sharp plasma transition since the accumulation of incorporated planetary ions to the solar wind through mass loading should proceed in a gradual manner [15]. Instead, it is necessary to consider that the transition detected downstream from the bow shock in Figure 4 derives from other phenomena, and since it defines the outer boundary of a region adjacent to the ionopause where lower values of the speed and density of the solar wind are measured, it should be related to transport of momentum to the Venus upper ionosphere, In particular, it has been suggested that such plasma transition represents the upper boundary of a region where solar wind momentum is transferred to the Venus ionosphere through viscous forces as it would occur when a streaming flow moves over an obstacle [10]. Its position should, therefore, indicate the width of a viscous boundary layer which increases with the downstream distances from Venus as it occurs in comparable fluid dynamic problems.

Further information on the presence of that plasma transitions in the Venus plasma environment has been obtained from spacecraft that have orbited the planet. This is the case for the Pioneer Venus and more recently for the Venus Express spacecraft whose data have strongly supported the presence of both the bow shock and the plasma transition located downstream from it [16-18]. A useful description of recent data is reproduced in Figure 6 from the Venus Express measurements whose trajectory, projected in cylindrical coordinates, is presented in Figure 7. The top and the second panels in Figure 6 show the

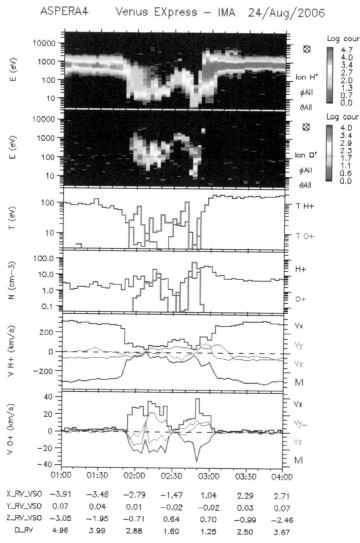

Fig. 6. Energy spectra of H+ and O+ plasma fluxes (first and second panels) measured in the Venus wake and over the north polar region during orbit 125 of the Venus Express (VEX) spacecraft in August 24, 2006. Temperature and density profiles of both plasma components (third and fourth panels) and their speed V, together width their Vx, Vy, and Vz, velocity components (fifth and sixth panels).

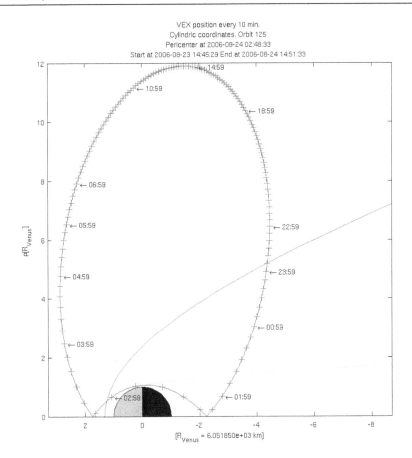

VEX position every 10 min.
Cylindric coordinates. Orbit 125
Pericenter at 2006-08-24 02:48:33
Start at 2006-08-23 14:45:29 End at 2006-08-24 14:51:33

Fig. 7. Trajectory of the Venus Express spacecraft in cylindrical coordinates around Venus during orbit 125 in August 24, 2006,[18].

energy spectra of the solar wind proton (H+) population and the planetary ions (mostly atomic oxygen O+) that were obtained as the spacecraft moved through the Venus wake in orbit 125 of August 24, 2006. Derived from those spectra are the temperature, density and velocity profiles of both ion components shown in the third, fourth, fifth and sixth panels of Figure 6 from which it is possible to see a bow shock crossing at 03:10 UT (the v_x velocity component in the sun-Venus direction and the speed value of the solar wind H+ ions are smaller in the downstream side, and their temperature and density are larger in that side indicating heating and a local compression). Evidence of a separate plasma transition can be identified at 01:52 UT and at 02:52 UT which bound a region where there are even smaller values in the speed of the H+ ion population and the observation of O+ ion fluxes (red trace). Related to this later plasma transition are the velocity components and the speed value of the planetary O+ ions shown in the bottom panel with an important piece of information; namely, their speed (20-30 km/s) is smaller than the local speed (50-150 km/s) of the solar wind H+ ions. Different speed values between both ion components would have not been expected if the planetary O+ ions were accelerated by the electric field that, as noted above, arises in the mass loading process (when exospheric neutral particles are ionized their speed is different from that of the solar wind). Rather than reaching solar wind speeds the planetary O+ ions maintain small speed values thus suggesting that their motion does not follow gyration around the magnetic field lines as it occurs when they enter a planetary magnetosphere. Instead, their response to local conditions is that they are slowly accelerated through processes that resemble those occurring in turbulent fluid dynamics. The data presented in Figure 6 thus indicates that the interaction of the solar wind with planetary ions in the Venus plasma environment can be viewed in the context of fluid dynamics with processes related to local turbulence and that provide the basis of that approach.

4.3 Plasma channels over the magnetic polar regions of the Venus ionosphere

Observations related to the geometry of the Venus nightside ionosphere and the plasma distribution in the wake are also supportive of a fluid dynamic view since they refer to well defined features that in many instances are present downstream from the planet. Such observations describe plasma clouds and ionospheric holes that were detected as the Pioneer Venus spacecraft moved through the nightside hemisphere within and in the vicinity of the ionosphere. As depicted in Figure 3 plasma clouds are regions that are separated from the main ionosphere and that in fact should be part of ionospheric streamers or filaments that extend downstream from Venus [19]. Ionospheric holes are, on the other hand, regions of very low plasma density that in some orbits of the PVO are detected as the spacecraft moves within the nightside ionosphere [20]. A large fraction of them are measured in the inner wake near the midnight plane but their location is displaced towards the dawn side. A profile of the electron density measured as the PVO moved through the nightside ionosphere in the PVO orbit 530 is shown in the upper panel of Figure 8. That profile describes the ionospheric holes that were detected at 09:30 UT and at 09:40 UT in the northern and in the southern hemisphere along the near polar oriented trajectory of the spacecraft (As it is shown in the lower panel of Figure 8 the lower values of the electron density within the holes are accompanied by enhanced magnetic field intensities) [21]. In some orbits the PVO only detected one ionospheric hole and in others none at al. Their origin and position have been the source of much research and, in particular, they have been interpreted as resulting from plasma channels (or ducts) that the solar wind produces as it moves over the magnetic polar regions of the Venus ionosphere [21]. A schematic view of such plasma channels is illustrated in Figure 9 to describe how they could be sometimes encountered along the PVO trajectory and thus be detected as ionospheric holes. The low density values of the ionospheric plasma within the plasma channels, and hence observed as ionospheric holes, could be due to the erosion that the solar wind produces upon reaching the (magnetic) polar regions. Favorable conditions for a turbulent interaction between the solar wind and the ionospheric plasma in those regions will arise since the local magnetic field intensity is not significantly enhanced as it occurs in the magnetic barrier that builds up around most of the dayside ionosphere [12].

The results of recent analysis of the Venus Express plasma data are also compatible with the presence of plasma channels that extend downstream from the polar regions and thus stress that further research is required to examine its geometric properties such as width, depth, and their evolution along the downstream distance from Venus. Independent of these characteristics it is to be noted that plasma channels in the polar ionosphere provide an important lead on the fluid dynamic interpretation of the data (plasma clouds,

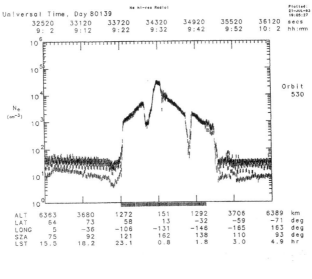

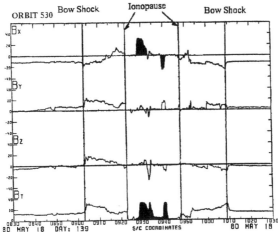

Fig. 8. (upper panel) Electron density profile measured across the nightside ionosphere in orbit 530 of the PVO. The ionospheric holes at 09:30 UT and at 09:40 UT were detected as the spacecraft moved through the northern (inbound pass) and southern (outbound pass) hemispheres [20]. (lower panel) Magnetic field components and magnetic field intensity B_T measured across the nightside ionosphere in orbit 530 (the shaded areas represent the position of the ionospheric holes) [21].

ionospheric holes). In particular, while much of the discussion presented here has been directed to describe those features the physical processes that produce them should still be addressed, namely, the manner in which turbulent flow conditions in a collisionless plasma lead to the fluid dynamic behavior.

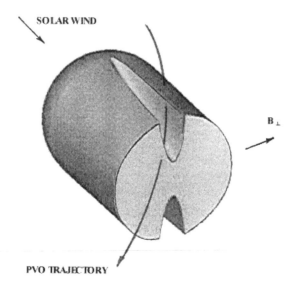

SOLAR WIND

$B_\perp$

PVO TRAJECTORY

Fig. 9. Schematic view of plasma channels that form by and behind the magnetic polar regions of the Venus Ionosphere [21].

4.4 Fluid dynamic Magnus force derived from the rotation motion of the Venus ionosphere

Unrelated to the erosion that the solar wind produces in the upper polar ionosphere there are other observations that also fit within the context of a fluid dynamic interpretation. These refer to a dawn-dusk asymmetry in the distribution of the ionospheric trans-terminator flow as it streams into the nightside hemisphere. Measurements conducted with the PVO show that the flow of ionospheric plasma behind the planet is not seen to converge towards the midnight plane but is oriented to the dawnside of the wake [22]. A general description of the manner in which the velocity vectors of the ionospheric flow are traced around and behind the planet (projected on the equatorial plane) is reproduced in the upper panel of Figure 10. As it was noted in Figure 5 the ionospheric flow is mostly measured by and downstream from the terminator but the distribution of the velocity vectors in the upper panel of Figure 10 indicate that the plasma experiences a dawn directed deviation at a ~15° angle away from the midnight plane. It is to be noted that the downstream displacement of the trans-terminator flow occurs mostly in the upper ionosphere (between ~ 400 km up to the ~ 1000 km altitude boundary of the ionosphere) and that it is superimposed on a different flow motion that carries the lower ionospheric plasma in the same retrograde rotation of the Venus atmosphere (measurements conducted with the PVO and the Venus Express spacecraft have shown that the neutral Venus atmosphere completes a steady rotation in a 4-5 earth day time span and that this motion is directed in the sense contrary to that of the earth´s rotation [23]). Figure 10 shows that the retrograde rotation of the Venus atmosphere is also applicable to the lower ionosphere and hence produces a velocity field that, as it is indicated in the dawn side, is superimposed on the trans-terminator flow. The kinetic energy of the latter flow provides an important contribution to drive the retrograde rotation of the atmosphere [24].

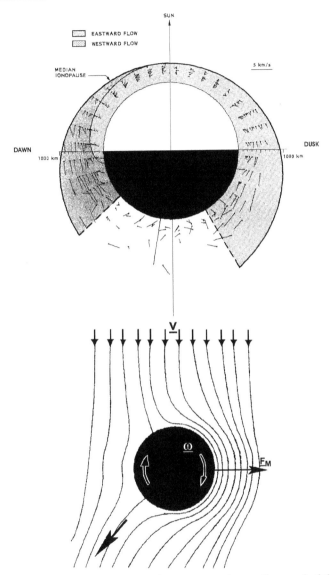

Fig. 10. (upper panel) Velocity averages of the trans-terminator flow and of the rotation of the low altitude ionosphere (which is better seen in the dawn side) that were measured in the Venus ionosphere with the PVO spacecraft and that are projected on the equatorial plane. The velocity vectors of the trans-terminator flow are directed away from the sun (the altitude scale has been increased four times in the figure) [22]. (lower panel) Streamline distribution around an obstacle rotating with a frequency ω immersed in a streaming fluid of velocity **V.** The Magnus Force F_M is directed transverse to the oncoming velocity direction and to the rotation vector ω that points out of the page. The arrow denotes the direction of the deviated wake [26].

The concurrent presence of the rotation motion of an object immersed in a directional flow as that inferred for Venus; namely, the nightward flow of the upper ionosphere and the rotational motion of the lower ionosphere is common in fluid dynamic problems and yields a deflection of the wake that can be used to account for the observations at Venus. A schematic description of the problem is illustrated in the lower panel of Figure 10 which reproduces conditions similar to those presented in the upper panel. The rotation motion of the obstacle produces a velocity field that yields velocity vectors parallel and anti-parallel to those of the directional motion. When they are parallel (right side in the lower panel of Figure 10) the added velocity is large, but becomes small when they are anti-parallel (left side). An important implication derived from the different speed values across the obstacle can be inferred from the Bernoulli´s equation:

$$P + \rho v^2 = cst \qquad (3)$$

which implies an inverse relation in the value of the pressure P across the obstacle (ρ is the mass density of the flow) and the kinetic energy of the flow ρv^2; that is, low pressure values should be expected in the region where the added flow speed is large (right side in the lower panel of Figure 10) and large pressure values will occur where the added speed is small (left side). In turn, the pressure difference across the obstacle derived from Bernoulli´s equation leads to a force (Magnus force) that is responsible for the lateral deviation of rotating baseballs and golf balls in sport activities (curve force) [25]. The dawnward directed deflection of the Venus trans-terminator flow indicated in the upper panel of Figure 10 results from the response force that the rotating Venus ionosphere applies on the trans-terminator flow (likewise the deflection of the wake in the lower panel of Figure 10 is produced in response to the Magnus force that the streaming flow applies on the rotating obstacle). Calculations were made to estimate the dawnward directed displacement of the trans-terminator flow produced by the Magnus force when using its 3-4 km/s flow speed and the rotation frequency of the lower ionosphere. The predicted displacement value is of the order of ~ 1000 km for the plasma located in the midnight ionosphere, which compares well with the measurements presented in the upper panel of Figure 10 (the dawnward deflection of the symmetry axis of the trans-terminator flow leads to that value at nightside ionospheric altitudes). In summary, the dawn-dusk asymmetry in the distribution of the trans-terminator flow around Venus is compatible with effects produced by the fluid dynamic Magnus force that applies to a rotating obstacle immersed in a directional flow [26].

4.5 Solar wind erosion of the Mars ionosphere

Much of the interaction of the solar wind with Mars leads to conditions that are similar to those observed at Venus. In the absence of a global intrinsic magnetization in both planets the solar wind reaches their atmosphere and produces phenomena that result in the erosion of their upper ionosphere. This view is applicable around most of Mars but it is modified in the vicinity of regions where there are fossil magnetic fields possibly remnants of an early global field [27] (in those regions the solar wind interacts with small scale magnetospheres of the type generated by the earth´s magnetic field). Conditions over the polar regions of the Mars ionosphere resemble those at Venus with ionospheric plasma being removed by the direct access of the solar wind. Information that supports the erosion of the Mars polar ionosphere was obtained from remote observations conducted with the XMM Newton satellite in orbit around the earth. A long time span image of Mars observed in the x-ray spectrum is presented in Figure 11 to describe the configuration of the region that has been eroded by the solar wind [28]. Most notable is the prominent plasma bulge that extends several thousand kilometers downstream from the Mars over the polar regions, and that is

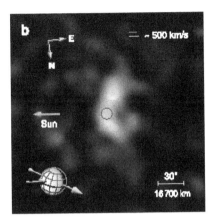

Fig. 11. View of the x-ray emission line halo around Mars measured with the reflecting grating spectrometer (RGS) of the XMM-Newton satellite. The halo is most prominent above the poles and is tilted away from the Sun. The sphere at the lower left indicates the direction of motion of Mars (green arrow) and the velocity of the solar wind particles (yellow arrow) [28].

similar to those occurring at the Venus polar regions. At Mars the solar wind carves out plasma channels with ionospheric plasma being continuously removed within a funnel cross-section that expands away from the wake. A comparable erosion of plasma also occurs by the polar regions of the Venus ionosphere as the channels are maintained open. However, because of the smaller gravitational force of Mars the density of the eroded plasma in that planet should be larger than at Venus, thus yielding evidence of the notable asymmetric plasma halo that is seen in its x-ray spectrum. The x-ray emission in the Mars spectrum corresponds to high energy electron transitions that imply very high temperatures. Calculations made from the emission line distribution in the x-ray Mars spectrum lead to temperature values of the order of one million degrees which may be produced through dissipation processes associated with the transport of solar wind momentum to the polar ionospheric plasma, as it was also suggested for Venus [29]. The remarkable shape of the Mars plasma halo shown in Figure 11 represents visible evidence of phenomena that substantiate the value of fluid dynamics in the interpretation of the solar wind interaction with planetary ionospheres. The collective removal of ionospheric plasma from the planets Venus and Mars through erosion processes is a useful byproduct of turbulent flow conditions that allow fluid dynamics to become applicable.

5. Concluding remarks

Much of what has been discussed in regard to the fluid dynamic view of the origin of the solar wind, its expansion through interplanetary space, and its interaction with the planets of the solar system leads to processes that are related to the motion of charged particles through magnetic fields and/or are subject to turbulent flow conditions. In the first case the solar wind particles enter the earth's magnetic field and produce electric currents that bound a cavity in which the magnetic field is compressed and is deformed to produce a magnetic tail. In the latter case the solar wind reaches planetary ionospheres where instead of gyration motion around the magnetic field lines the plasma particles describe stochastic trajectories produced by local turbulence and that suggest the existence of wave-particle interactions. The outcome of this latter activity is the onset of phenomena that can be interpreted in terms of those occurring in fluid dynamic problems. A process that has been applied in this context is the transport of solar wind momentum that occurs along the flanks of planetary ionospheres as it is shown in Figure 4 (the boundary traced between events 2 and 4 represents the outer extent of a viscous layer). For those conditions it has been possible to estimate the viscosity coefficient of the solar wind that is suitable to account for the observations. The calculated values of the kinematic viscosity of the solar wind suggest that they are compatible with the plasma properties of the solar wind; namely, an efficient transport of momentum across a thick ($\sim$ 10^3 km) viscous boundary layer measured by the terminator and that is produced under its very low particle densities. In fact, since the value of the kinematic viscosity coefficient v describes the ability of a flow to modify a velocity profile [30] the small density of the solar wind leads to large $v \sim 10^4$ km^2/s values when compared with those of comparable fluid dynamic problems [31]. Empirical values of the kinematic viscosity of the solar wind are derived on the basis that local turbulence produced by wave-particle interactions provide the origin of the fluid dynamic interpretation [7, 32]. However, while there are various phenomena that substantiate this view it is still necessary to determine the mechanisms that produce such interactions which lead to the collective response of the solar wind. A required source of information would be obtained from a statistical mechanics applicable to wave-particle interactions and that could provide in mathematical form expressions for the transport coefficients. Much theoretical research should be conducted to improve the information that is now available.

6. References

[1] Parker, E. N., Interplanetary Dynamic Processes, John Wiley, 1963.

[2] Liepmann, H. W. and A. Roshko, Elements of Gas dynamics, John Wiley, 1967 (Chapter 2, p. 52).

[3] Dessler, A. J., Solar Wind and Interplanetary Magnetic Field, Reviews of Geophysics. 5,1, 1967.

[4] Spreiter, J., et al., Hydromagnetic flow around the magnetopause, Planetary Space Sciences, 14, 223, 1966.

[5] Dryer, M., et al., Magnetogasdynamic conditions for a closed magnetopause, American Institute Aeronautics Astronautics Journal (AIAA)., 4, 246, 1966.

[6] Brace, L., et al., The Ionotail of Venus: Its configuration and evidence for ion escape, Journal of Geophysical Research, 92, 15, 1987.

[7] Bridge H., et al., Plasma and magnetic fields observed near Venus, Science, 158, 1669, 1967.

[8] Knudsen, W., et al., Improved Venus ionopause: Comparison with measurements, Journal of Geophysical Research, 87, 2246, 1982.

[9] Whitten, R. et al., Dynamics of the Venus ionosphere: A 2D model study, ICARUS, 60, 317, 1984.

[10] Pérez-de-Tejada, H., Fluid dynamic constraints of the Venus ionospheric flow, Journal of Geophysical Research, 91, 6765, 1986.

[11] Pérez-de-Tejada, H. Friction layer in the plasma channels of the Venus ionosphere, Advances of Space Research, 36, 2030, 2005.

[12] Zhang, T., J. Luhmann, and C. Russell, The magnetic barrier at Venus, Journal of Geophysical Research, 96, 11145, 1991.

[13] Pérez-de-Tejada, H., et al., Plasma distribution in the Venus near wake, Journal of Geophysical Research, 88, 9109, 1983.

[14] Breus, T. et al., Solar wind mass-loading at comet Halley, Geophysical Research Letters, 14, 499, 1987.

[15] Intriligator, D., Observation of mass addition in the Venusian ionosheath, Geophysical Research Letters, 9, 727, 1982.

[16] Slavin, J., et al., The solar wind interaction with Venus: PVO-Venus bow shock, Journal of Geophysical Research, 85, 7625, 1980.

[17] Pérez-de-Tejada, H., et al., Intermediate transition of the Venus ionosheath, Journal of Geophysical Research, 100, 14523, 1995.

[18] Pérez-de-Tejada, H. et al, Plasma transition at the flanks of the Venus ionosheat: Evidence from the Venus Express, Journal of Geophysical Research, 116, A01103, 2011.

[19] Brace , L, et al., Plasma clouds above the ionosphere of Venus and their implications Planetary Space Sciences, 30, 29, 1982.

[20] Brace, L., et al., Holes in the nightside ionosphere of Venus, Journal of Geophysical Research, 87, 199, 1982.

[21] Pérez-de-Tejada, H., Plasma channels in the Venus nightside ionosphere, Journal of Geophysical Research, 109, A04106, 2004.

[22] Miller, K., and R. Whitten, Ion dynamics in the Venus ionosphere, Space Science Reviews, 55, 165, 1991.

[23] Schubert, G., et al., Structure and circulation of the Venus atmosphere, Journal of Geophysical Research, 85, 8007, 1980.

[24] Lundin, R. et al., Ion flow and momentum transfer in the Venus environment, ICARUS (in press, 2011).

[25] Rouse, H., Elementary Mechanics of Fluids, chap. IX, Dover, Mineola, N.Y. 1978.

[26] Pérez-de-Tejada, H., The Magnus force in the Venus ionosphere, Journal of Geophysical Research, 111, A11105, 2006.

[27] Acuña, M., et al., Global distribution of crustal magnetization at Mars, Science, 184, (5357), 790, 1999.

[28] Dennerl, K., et al., First observation of Mars with XMM-Newton, Astronomy and Astrophysics, 451, 709, 2006.

[29] Pérez-de-Tejada, H. et al, Solar wind erosion of the Mars polar ionosphere, Journal of Geophysical Research, 114, A02106, 2009.

[30] Batchelor, G. K. An Introduction to Fluid Dynamics, Cambridge University Press (p. 36), 1979.

[31] Pérez-de-Tejada, H., Viscous forces in a boundary layer at the Venus ionosphere, Astrophysical Journal, 525, L65, 1999.

[32] Vörös, Z., et al., Intermittent turbulence, noisy fluctuations, and wavy structures in the Venusian magnetosheath and wake, J. Geophys. Res.-Planets, 113, E00B21, doi:10.1029/2008, 2008.

10

Aero - Optics: Controlling Light with Air

Cosmas Mafusire[1,2] and Andrew Forbes[1,2]

[1]Council for Industrial Research National Laser Centre
[2]School of Physics, University of KwaZulu-Natal
South Africa

1. Introduction

It is possible to control light in a medium comprising nothing more than a fluid, whether in gas or liquid phase. Using air as an example, it is possible to bend light by a process of continuous refraction as opposed to stepwise refraction in the case of solid state optics. This is achieved by effecting a continuous change of the refractive index from a high value for lowest temperature (or high density) to a low value for the highest temperature (or lowest density). Since the density changes gradually from a low to a high value, the refractive index is graded as well. In fact, this effect is well known in nature: the mirage effect where light refracts away from a hot surface, usually a road or desert sand, is precisely due to continuous refraction of light. If the refractive index can be customised and controlled, then one has the possibility to create aero-optics: control of light with air. Such optics are a special subset of GRIN (Graded Refractive INdex) optics, with the most studied example being the simple lens, so-called gas lenses. The efficiency at which these lenses operate is dependent on the effectiveness of the refractive index gradient – the greater the gradient, the stronger the lens. There are advantages to such optics: they are not dispersive, and hence suitable for broad bandwidth laser pulses, and have no practical damage threshold as compared to most solid-state optics (e.g., glass). This latter point remains to be exploited for high power laser beam delivery in such applications as laser fusion and peta-watt high intensity laser beam delivery.

In the remainder of this chapter we will consider such gas lenses in more detail. We outline a simple approach to controlling light through a medium comprising nothing more than air in a spinning pipe that is heated along its boundary. We outline the fluid dynamics of this simple system and through the use of computational fluid dynamics (CFD), explain the mechanisms that give rise to a plethora of interesting properties, from a lensing action resulting in the focussing of light, to a simulator of optical turbulence in the laboratory. The CFD models are supported in most cases by intuitive analytical expressions, teasing out fundamental physics from the fluid dynamics of this system, while the predicted impact of the fluid dynamics on the optical field is verified experimentally. Despite the simplicity of the system – merely a spinning, heated pipe – the fluid dynamics of the system is shown to be rather complex, comprising regions of slow and fast flow, symmetric and asymmetric density gradients, and laminar and turbulence regions. We conclude the study by outlining how such aero-optics, using aerodynamics and fluid flow to control light, may be exploited in applications ranging from long range telescopes to overcoming the damage thresholds of conventional solid state glass optics for delivery of high power lasers.

2. Gas lenses

The principle of operation of the spinning, heated, pipe as a gas lens, an example of which is shown in Fig. 1, is based on the concept that the shear viscosity of air increases when heated, so that when the outside wall of the pipe is heated the layer of air next to the pipe's inner wall is heated as well. During rotation this inner layer is centrifugally expelled and replaced by cold air from the surroundings, inhaled along the axis from both ends, resulting in an axial (air) region that is cooler compared to the air at the wall. This temperature gradient creates a density gradient resulting in a refractive index gradient conducive to focusing, thus the name: gas lens. Rotating the heated pipe converts the density with a vertical gradient due to gravity to one that is rotationally symmetric about the axis. In order to customise the aero-optic for a lensing action, it is necessary to create a parabolic density gradient (and hence temperature gradient) across the pipe, and preferably everywhere along its length. A purely parabolic gradient in the density would be an ideal GRIN lens: the perfect gas lens.

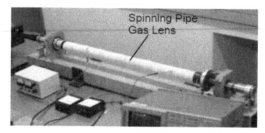

Fig. 1. An image of the spinning pipe gas lens (SPGL), comprising a steel pipe that is heated and spun around its axis. The central heated section is evident by the white heater tape surrounding it.

Previous studies have shown that the SPGL imitates a graded index (GRIN) medium with a refractive index which is a maximum along the axis and decreases parabolically with radial distance r, towards the walls (Michaelis, 1986; Forbes, 1997; Mafusire, 2006)

$$n(r) = n_0 - \tfrac{1}{2}\gamma^2 r^2 \tag{1}$$

The radial refractive index parameter, γ is a measure of the power of the SPGL, while n_o is the refractive index along the axis.

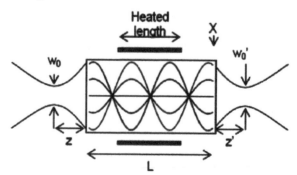

Fig. 2. A laser beam being converged on exit from a spinning pipe gas lens.

The focal length of the lens in terms of the input and output beam waist positions, w_0 and w_0', located at respective distances, z and z' with respect to the SPGL (Fig. 2) is given by (Mafusire, 2008b)

$$f = \frac{1}{2(z'-z)}\left[z_0^2 + \frac{1}{\gamma^2} + 2zz' - z \pm \sqrt{z_0^4 + 2z_0^2\left(\frac{1}{\gamma^2} + 2z' - z^2\right) + \left(\frac{1}{\gamma^2} + z^2\right)^2} \right] \qquad (2)$$

where z_0 is the beam's Rayleigh range in front of the lens.

The earliest gas lenses were made at Bell Laboratories in an early attempt at producing waveguides for long range communication as a forerunner to the optical fibre. They were operated by heating the walls to a suitably high temperature then cold air was injected along the axis, creating a graded index between the lower density of the hot marginal air and increasing radially towards the cooler axis. Any laser beam propagating along the axis is refracted towards the denser axis hence were referred to as the tubular gas lenses. Early work was done by Berreman (Berreman, 1965) and Marcuse (Marcuse, 1965). In fact Marcuse showed that these lenses approximate a thin lens since both their principal planes coincide. Beyond the potential use for waveguiding (Kaiser, 1968), they were also used as telescope objectives (Aoki, 1967).

The main disadvantage with this design was that convection currents were clearly present and so introduced aberrations due to gravity on the laser beam wavefront. Gloge (Gloge, 1967) showed that the optical centre of such lenses is displaced vertically downwards, an effect which increased with tube diameter. Kaiser (Kaiser, 1970) circumvented this problem by designing a gas lens where hot gas could be exhausted radially, with the cold gas injected as before, resulting in improved performance. Meanwhile, in the former Soviet Union, another improvement was made by replacing the gas injection with rotating a heated pipe about its axis (Martynenko, 1975). It was demonstrated, theoretically and experimentally, that the rotation removed distortions due to gravity by creating a rotationally symmetric density distribution which completely overcame the convection currents of the tubular gas lens.

The investigation into this version of this gas lens design, now called the spinning pipe gas lens, or SPGL for short, was taken over by the then University of Natal without the knowledge of the research in the former USSR. They worked with a vertical lens which was used for focusing a high power laser for drilling holes into metal sheets (Michaelis, 1986). Another improvement was to combine both rotation and gas injection to create a steady focus (Notcutt, 1988). An interesting application was when the SPGL was used as a high quality telescope objective to take images of sun spots and moon craters, illustrating the power of such devices (Michaelis, 1991). Further improvements included operating the SPGL at pressures higher and lower than atmospheric pressure (Forbes, 1997) in order to control the focal length, and careful characterisation of the temperature distribution inside the pipe (Lisi, 1994).

Recently a more modern approach to understanding such devices has been completed, considering the interface of fluid dynamics with physical optics propagation and characterisation of laser beams (Mafusire et al, 2007; Mafusire et al, 2008a; Mafusire et al, 2008b; Mafusire et al, 2010a; Mafusire et al, 2010b). Using computational fluid dynamics (CFD) to simulate the density and velocity distributions inside the SPGL, one can deduce the phase change imparted to the light through the Gladstone-Dale law, thus making it possible to calculate optical aberrations at any plane along the beam's path. To appreciate this, consider the gas medium where the density at each point in the medium fluctuates about a certain mean density, σ. These density fluctuations result in refractive index fluctuations also about a mean refractive index, N. These two parameters are related by the Gladstone-Dale law given by (Zhao et al, 2010)

$$N = G(\lambda)\sigma + 1 \qquad (3)$$

where $G(\lambda) = 2.2244 \times 10^{-4}[1 + (6.7132 \times 10^{-8} / \lambda)^2]$, the wavelength dependant Gladstone -Dale constant which has a value of about 2.25×10^{-4} m³/kg in air for HeNe laser radiation ($\lambda = 6.328 \times 10^{-7}$ m). If the geometrical length of the laser beam path in the gas lens is given by, l, then the phase change of the laser beam is given by

$$\phi = klN \qquad (4)$$

where $k = 2\pi / \lambda$ is the laser beam wavenumber for radiation of wavelength, λ. The implication is that the optical distortions in the gas lens, due to imperfect control of the fluid to create the lens, can be measured by observing the phase changes in the laser beam. Distortions to the phase are called optical aberrations. The measurement of the optical aberrations of this particular aero-optic has introduced a very interesting paradox: rotation of the gas lens removes distortions due to gravity, but as we will see later, the other aberrations increase in magnitude as the rotation speed and the wall temperature increase. Therefore such optics are not perfect. To understand where the imperfections originate from, we start with a computational fluid dynamics (CFD) model of the gas lens.

2.1 Computational fluid dynamics model of the gas lens

To confirm the theoretical analysis of the SPGL given above and the optical aberrations measured from its performance, a computational fluid dynamics (CFD) simulation of a simplified test system was executed using the commercial CFD code, STAR–CD® using the $k - \varepsilon$ model which involves the numerical solution of two coupled equations, a turbulent kinetic energy (k) equation and an energy dissipation rate (ε) equation [Blazek, 2001; Davison, 2011]. The purpose of this study was to show the effect of the heat and mass transfer on the velocity distribution and density. Assumptions included the removal of the mounts and other three-dimensional geometry features that would complicate the geometric model. The pipe used in the model was based on the dimensions of the actual SPGL we have in the laboratory: 1.43 m in length and with an internal diameter of 0.0366 m. The heated section of the pipe was approximately 0.91 m long, leaving two unheated end lengths of length 0.25 m each. The pipe was accurately reproduced with the further assumption that the mounts act as a heat sink and thus the pipe ends are unheated. A fully transient solution is presented in which the pipe is spun up from a heated (at a temperature of 100 °C) steady-state buoyancy-driven solution, and held at fixed speed of 20 Hz until a steady state was reached.

The mesh used in the solution consisted of 350 transverse slices regularly arranged along the pipe's length. Each slice consisted of 512 nodes distributed across the section with more nodes concentrated along the boundary (see Fig. 3). The results extracted from the model included density for each data point in the centre of the cell in the mesh and animations showing the evolution of the velocity and density at selected transverse and longitudinal cross-sections. Temperature distribution was extracted from the density data.

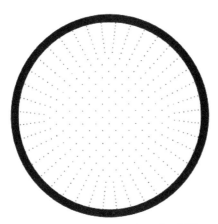

Fig. 3. The SPGL mesh transverse cross-section used in the CFD model.

2.1.1 Velocity distribution

Fig. 4 is a longitudinal cross-section of the gas lens showing the velocity distribution for various cuts of the pipe: (a) left third of the pipe, (b) the mid-section of the pipe, and (c) the right third of the pipe. The velocity distribution in the unheated sections is dominated by transverse movement of air towards the centre from both sides in the inviscid section

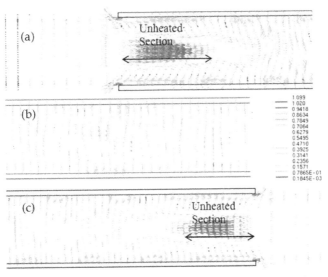

Fig. 4. Velocity profiles in the SPGL for the (a) left end section, (b) the mid-section of the pipe, and (c) the right end section.

outwards along the boundary section, with rotation playing very little or no part. As one moves towards the centre of the pipe length, the transverse speed of incoming air decays (as expected). In the boundary layer, the expelled air is fastest as it exits from the pipe. As the air along the boundary approaches the pipe end, it increases in both translation and rotational velocity, thus the boundary expels hot air in a spiral motion at both ends of the pipe. From this velocity distribution, we can infer the heat distribution in the inviscid region since fast moving air accumulates transversely injected heat much more slowly compared to stationary air. This means the incoming air is cooler but accumulates heat as it slows down and approaches the centre. This might point to a situation where the temperature along the axis at the centre is higher than that of the walls: because of the particles' slow velocity, they accumulate heat over time.

A closer look at the velocity profile of Fig. 4 (b) shows the complicated velocity distribution away from ends of the pipe. This could be evidence of multi-cellular flow, the irregularity of which is most likely responsible for the presence of some of the optical aberrations. This oscillatory activity is expected to increase with increase in wall temperature or rotation speed, which in turn, increases the magnitude of the optical aberrations.

2.1.2 Density distribution

The initial state of the gas in a heated stationary heated SPGL is a result of natural convection. A CFD simulation of this is shown in Fig. 5 (a): it shows a density gradient that decreases vertically so that at the gas is layered in horizontal bands within the pipe. When the pipe is spun the image changes radically: the vertical density gradient gives way to a density gradient that has rotational symmetry due to the forced convection of the rotating system, with the highest density along the axis and decreases towards the edges of the pipe (see Fig. 5 (b)). This is an example of a customised aero-optic: such a density distribution is conducive to focusing, since it results in a refractive index profile given by Eq. (1).

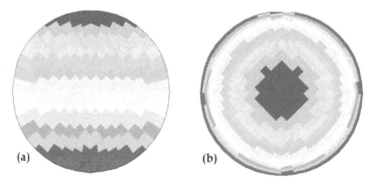

(a)   (b)

Fig. 5. Cross–sectional density profiles of an SPGL showing: (a) the initial state after heating, and (b) the rotating steady–state near the end face of the pipe, with high density centre (blue) and low density edges (red).

The other results from the density calculations confirm the velocity profile from the previous sub-section. The temperature (T) distribution was extracted from the density (ρ) data using the relation, $\rho = \rho_0(1 - a(T - T_0))$ where a = 2.263123×10⁻⁴ K⁻¹ is the coefficient of volume expansion of air and ρ_0 is the known density at a known temperature, T_0 which we take to be room temperature. Fig. 6 (a) shows a 3D plot of the density profile of the air everywhere along the length of the pipe for a particular (but arbitrary) cross-section. From this one can visualise the density distributions in transverse sections at critical planes, and a zoomed in view of the centre of the pipe is show in Fig. 6 (b). The corresponding temperature profile shows that the unheated section is at room temperature (27 °C) and the centre of the pipe is at about 100 °C, the temperature of the wall.

From Figs. 4 and 6 we can start to understand how this particular device works: the intake of cold and the expulsion of hot air, which takes place the ends of the pipe, results in parabolic-like density profiles near the pipe ends, which results in the lensing effect. As can be observed in the transverse profiles, a significant parabolic density distribution is only evident in a short section of the entire length, possibly less than half. As the pipe is spun faster, so this "lensing length" increases, making the lens stronger. A closer look at the cross-section profiles in Fig 6 (a) shows that some of them have a quartic density distribution, albeit for a very short length compared to the purely parabolic density profile, and this is the source of spherical aberration which has been shown to increase with rotation speed and/or temperature (Mafusire et al, 2008a).

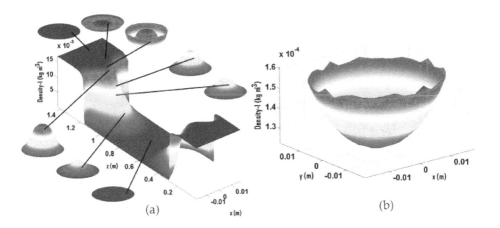

(a)   (b)

Fig. 6. (a) A 3D longitudinal density profile for the SPGL showing transverse sections at critical planes, and (b) the transverse density profile at the centre of the SPGL.

Another interesting observation from the CFD model of the SPGL is the transverse temperature profile in the centre of the pipe, i.e. away from the face ends, which reveals that the temperature along the axis is just over 0.16 K higher than at the walls, corresponding to a density 4×10^{-5} kg m^{-3} lower than at the walls (Fig 6 (b)). The explanation for this is that because air particles at this part of the pipe are moving very slowly and thus accumulates heat faster due to both convection and conduction. Thus, somewhat counter intuitively, the central heated part of the gas lens actually results in a decrease in the lensing action.

2.2 Experimental verification of the gas lensing
The experimental setup used is shown in Fig. 7. When the pipe is not rotated and not heated, no lensing takes place, and the phase of the light is taken as a reference. One expects

that if the pipe acts as a lens, the parabolic refractive index will result in curvature on the phase of the light, indicative of a lens. To test this, the pipe was heated to wall temperatures of 348 K, 373 K, 398 K and 423 K, and spun at various rotation speeds (limited to 20 Hz in our experiments). As the pipe is spun, so the focal length of the lens decreased, suggesting a stronger lens. Similarly, when the wall temperature of the pipe was increased, the lensing became stronger. Experimental results of this lensing is shown in Fig. 8.

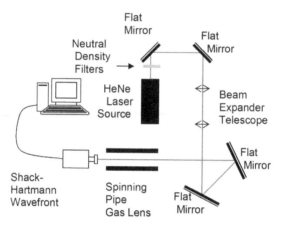

Fig. 7. Experimental setup used to test the lensing properties of the gas lens.

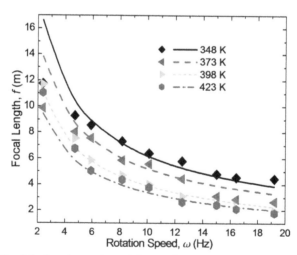

Fig. 8. The strength of the lensing action of the gas lens increases with both pipe rotation speed and pipe wall temperature. Short focal lengths suggest a strong lens, while long focal lengths suggest a weak lens.

3. Optical aberrations

We have seen earlier that the gas lens is aberrated due to the inability to completely control the fluid as desired. In order to describe just how aberrated the lens is, we introduce the concept of optical aberrations, and their description by Zernike polynomials, the coefficients of which are correlated to the magnitude of the optical aberrations. Once these coefficients are known, we can calculate the rms phase error (the deviation from flatness) and the Strehl ratio – how good the focus is for retaining power in a small spot.

3.1 Zernike polynomials

The phase of a laser beam can be written as a linear combination of Zernike polynomials. Zernike polynomials are unique in that they are the only polynomial system in cylindrical co-ordinates which are orthogonal over a unit circular aperture, invariant in form with respect to rotation of the co-ordinate system axis about the origin and include a polynomial for each set of radial and azimuthal orders [Born & Wolf, 1998; Dai, 2008; Mahajan, 1998; Mahajan, 2001]. More importantly, the coefficients of these polynomials can be directly related to the known aberrations of laser beams, making them invaluable in the description of phase errors. The transverse electric field of a laser beam, U, in cylindrical coordinates, ρ and θ can be represented by the product of the amplitude ψ and phase φ as shown by

$$U(\rho,\theta) = \psi(\rho)e^{i\varphi(\rho,\theta)} \tag{5}$$

The expansion of an arbitrary phase function, $\varphi(\rho,\theta)$, where $\rho \in [0,1]$ and $\theta \in [0, 2\pi]$, in an infinite series of these polynomials will be complete. The circle polynomials of Zernike have the form of an angular function modulated by a real radial polynomial. We can represent each Zernike term by

$$Z_{nm}(\rho,\theta) = C_{nm}R_{nm}(\rho)\Theta_m(\theta) \tag{6}$$

The angular part is defined as

$$\Theta_m(\theta) = \begin{cases} cosm\theta, & m > 0 \\ sinm\theta, & m < 0 \\ 1, & m = 0 \end{cases} \tag{7}$$

whereas the radial part is a polynomial given by

$$R_{nm}(\rho) = \sum_{k=0}^{\frac{n-m}{2}} \frac{(-1)^k (n-k)! \rho^{n-2k}}{k!(\frac{n+m}{2} - k)!(\frac{n-m}{2} - k)!} \tag{8}$$

where n and m are the non-negative order and ordinal numbers respectively which are related such that $m \leq n$ and $n - m$ is even. C_{nm} is the respective coefficient for a particular aberration and can be either A_{nm} and B_{nm} depending whether the aberration is either even or odd, respectively.

This means that a laser beam wavefront described by a phase function φ can be expanded as a linear combination of an infinite number of Zernike polynomials, using generalized coefficients as follows

$$\varphi(\rho,\theta)=2\pi\sum_{n=0}^{\infty}A_{n0}R_{n0}(\rho)+2\pi\sum_{n=1}^{\infty}\sum_{m=1}^{n}R_{nm}(\rho)[A_{nm}cosm\theta+B_{nm}sinm\theta] \qquad (9)$$

n	m						
	-3	-2	-1	0	1	2	3
0							
1							
2							
3							
4							

Table 1. Contour plots of the Zernike primary aberration polynomials.

n	m	Description and symbol	Polynomial
0	0	Piston, A_{00}	1
1	-1	y-Tilt, B_{11}	$\sqrt{2}\rho\sin\theta$
	1	x-Tilt, A_{11}	$\sqrt{2}\rho cos\theta$
2	-2	y-Astigmatism, B_{22}	$\sqrt{6}\rho^2 sin2\theta$
	0	Defocus, A_{20}	$\sqrt{3}(2\rho^2-1)$
	2	x-Astigmatism, A_{22}	$\sqrt{6}\rho^2 cos2\theta$
3	-3	y-Triangular Astigmatism, B_{33}	$\sqrt{8}\rho^3 sin3\theta$
	-1	y-Primary Coma, B_{31}	$\sqrt{8}(3\rho^3-2\rho)sin\theta$
	1	x-Primary Coma, A_{31}	$\sqrt{8}(3\rho^3-2\rho)cos\theta$
	3	x-Triangular Astigmatism, A_{33}	$\sqrt{8}\rho^3 cos3\theta$
4	0	Spherical Aberration, A_{40}	$\sqrt{5}(6\rho^4-6\rho^2+1)$

Table 2. The names of the Zernike primary aberration coefficients.

The A and B terms are referred to as the symmetric and non-symmetric coefficients, respectively, in the units of waves. If the phase is known as a function, then the *rms* Zernike coefficients A_{nm} and B_{nm} can are calculated using

$$A_{nm}=\frac{1}{\pi}\sqrt{\frac{2(n+1)}{1+\delta_{m0}}}\int_0^{2\pi}\int_0^1\varphi(\rho,\theta)R_{nm}(\rho)cosm\theta\rho d\rho d\theta \qquad (10a)$$

$$B_{nm} = \frac{1}{\pi} \sqrt{\frac{2(n+1)}{1+\delta_{m0}}} \int_0^{2\pi} \int_0^1 \varphi(\rho,\theta) R_{nm}(\rho) \sin m\theta \rho d\rho \theta \qquad (10b)$$

where δ_{m0} is the Kronecker delta function and the integrals' coefficients are normalizing constants. The names of the coefficients of primary aberrations which we are going to discuss in this paper are given in Table 1. From the aberration coefficients, we can calculate the wavefront error, the Strehl ratio (Mahajan, 2005), focal length (Mafusire and Forbes, 2011b) and the beam quality factor (Mafusire & Forbes, 2011a). The Strehl ratio is defined as the ratio of the maximum axial irradiance of an aberrated beam over that of a diffraction limited beam with the same aperture size. The general definition of the Strehl ratio is given by (Mahajan, 2001; Mahajan, 2005).

$$S = \left(\frac{\left| \int_0^{2\pi} \int_0^1 \psi(\rho) e^{i\phi(\rho,\theta)} \rho d\rho d\theta \right|}{\int_0^{2\pi} \int_0^1 \psi(\rho) \rho d\rho d\theta} \right)^2 \qquad (11)$$

$$\approx e^{-(\Delta\varphi^2)}$$

In the last result, we have used an approximation of the Strehl ratio for small aberrations less than the wavelength of the radiation. For a better undstanging on the wavefront error and Strehl ratio, the interested reader is referred to Mahajan (Mahajan, 2001; Mahajan, 2005).

3.2 Calculation of Zernike primary aberrations from CFD density data

A practical way to extract Zernike coefficients from CFD density data is to generate phase data in layers normal to the propagation direction of the beam using Eq. 4 with distance l between layers. A Taylor polynomial fit given by Eq. 12 is made to each plane to create a phase function in Cartesian form.

$$\begin{aligned} \varphi(x,y) = &p_{00} + p_{10}x + p_{01}y + p_{02}y^2 + p_{11}xy + p_{20}x^2 + p_{03}y^3 + p_{12}xy^2 + p_{21}x^2y \\ &+ p_{30}x^3 + p_{04}y^4 + p_{13}xy^3 + p_{22}x^2y^2 + p_{31}x^3y + p_{40}x^4 + p_{05}y^5 + p_{14}xy^4 \\ &+ p_{23}x^2y^3 + p_{32}x^3y^2 + p_{41}x^4y + p_{50}y^5 \end{aligned} \qquad (12)$$

We can then convert the Cartesian coordinates to cylindrical coordinates using $x = r\cos\theta$ and $y = r\sin\theta$, replace r with ρ/a then reduce the resultant function. If we substitute this function into Eq. 10 we can then extract the *rms* primary Zernike coefficients which are given by Eq. 13. These equations enable us to calculate the Zernike coefficients from the phase distribution data. With these data, it is possible to work out the phase change experienced by the laser beam as it is propagates through each layer. If we have data for the next layer, and the one after that, then it is possible to work out the impact on the laser beam as the output from one layer becomes the input for the next layer. If the density of each layer is known from a CFD model, then propagating the laser beam through the medium in this manner will approximate the total aberrations imparted to the laser beam, assuming the propagation distance is small. In our simulations we have restricted our description of the aberrations to the fourth order, but of course one can make the expansion as accurately as one desires.

$$A_{00} = \frac{1}{24}(a^2(a^2(3p_{04} + p_{22} + 3p_{40}) + 6p_{02} + 6p_{20}) + 24p_{00})$$

$$A_{11} = \frac{1}{96} a (a^2(3 (a^2(p_{14} + p_{32} + 5p_{50}) + 8p_{30}) + 8p_{12}) + 48p_{10})$$

$$B_{11} = \frac{1}{96} a (a^2(3 a^2(5p_{05} + p_{23} + p_{41}) + 24p_{03} + 8p_{21}) + 48p_{01})$$

$$A_{20} = \frac{1}{16\sqrt{3}}(a^2(a^2(3p_{04} + p_{22} + 3p_{40}) + 4p_{02} + 4p_{20}))$$

$$A_{22} = \frac{1}{8\sqrt{6}}(a^2(-3a^2(p_{04} - p_{40}) - 4p_{02} + 4p_{20}))$$

$$B_{22} = \frac{1}{16\sqrt{6}}(a^2(3a^2(p_{13} + p_{31}) + 8p_{11})) \tag{13}$$

$$A_{31} = \frac{1}{120\sqrt{2}}(a^3(3 (a^2(p_{14} + p_{32} + 5p_{50}) + 5p_{30}) + 5p_{12}))$$

$$B_{31} = \frac{1}{120\sqrt{2}}(a^3(3a^2(5p_{05} + p_{23} + p_{41}) + 15p_{03} + 5p_{21}))$$

$$A_{33} = -\frac{1}{40\sqrt{2}}((a^3(a^2(3p_{14} + p_{32} - 5p_{50}) + 5p_{12} - 5p_{30}))$$

$$B_{33} = \frac{1}{40\sqrt{2}}(a^3(a^2(-5p_{05} + p_{23} + 3p_{41}) - 5p_{03} + 5p_{21}))$$

$$A_{40} = \frac{1}{48\sqrt{5}}(a^4(3p_{04} + p_{22} + 3p_{40}))$$

3.3 Piston as a measure of average density

It is well known that piston is the average phase of a wavefront (Mahajan, 1998; Dai, 2008). In our formulation, Zernike coefficients are in the units of waves and phase is in radians, so that the piston is related to the average phase by $A_{00} = \frac{1}{2\pi}\bar{\varphi}$. Substituting for the phase as defined in Eq. 4 and introducing a new term, $L = l / \lambda$, the piston becomes $\bar{\varphi} = L\bar{N}$ where $\bar{N}$ is the average refractive index. Finally, combining these two equations and then substituting for the average refractive index as defined by Eq. 3, the result is

$$A_{00} = L(G\bar{\sigma} + 1) \tag{14}$$

The equation implies that the piston measured for a beam having passed through a medium of length, l, is directly proportional to average density in that medium. This means that if you have divided a propagation path of a laser beam in an aero-optic medium to steps each of length, l, then the variation of piston for each step is actually the variation in average density. This equation can also mean that if the piston is known after the beam has passed through such a medium, then we can use it to calculate the total average density of that path. If the density in the path is a constant, or $\sigma = \bar{\sigma}$, measuring piston can now be used to measure density. This would be true if all aberrations, except for piston, are zero, or at the very least, very small compared to piston. We can also conclude that the greater piston is compared to other aberrations, the more uniform the medium is.

3.4 Extraction of the optical parameters from the CFD model

Returning to our gas lens, CFD density data was extracted from 351 planes equally spaced along the length of the pipe, so that we model the pipe as 350 layers, or "individual gas lenses" placed one after another, each of length 4.1 mm. Density data is converted to refractive index data using the Gladstone-Dale's law and from this, the phase, φ is calculated using Eq. 4 with l set to 4.1 mm. A Taylor polynomial surface fit was carried out for each plane to create a phase function in Cartesian form. The coefficients thus acquired were used to calculate the Zernike coefficients, using Eq. 13, resulting in 350 sets of primary Zernike coefficients for the spinning pipe gas lens. It is instructive to partition the SPGL data into sections based on the known fluid behaviour discussed earlier, as illustrated in Fig. 9. These are the end sections, labelled A, which are just over 20 cm each, the central section, C, of about 20 cm. This leaves the two sections in-between, B, each about 40 cm. Sections A are unheated. Let us assume that the laser beam is propagating through the pipe with its axis coinciding with that of the pipe. The beam should be small enough not to experience any diffraction with the pipe walls. For that reason we choose a, of size 0.371 cm against a pipe of radius 1.83 cm. The size of the beam was set at 0.548 cm, the same beam size we used as in the experiment (Mafusire, 2008a). We further assume that the only aberrations the beam experiences are from the medium and not from diffraction due to the propagation itself.

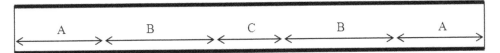

Fig. 9. Sections for analysing the spinning pipe gas lens based on the fluid behaviour disussed earlier.

We now present the graphs of all the primary aberrations as shown in Fig. 10. The aberrations have been organised starting with piston, followed by defocus, spherical aberration, tilt, coma and then astigmatic aberrations, i.e., astigmatism and triangular astigmatism (see Figure 10 (a)-(e)). Piston, Fig. 10 (a), has a characteristic curve which shows local average density, related to the overall phase delay experienced by the beam (Eq. 14). It is maximum in section A of the gas lens and minimum in section C. Section A is where the pipe is not heated and section C is the hot section where rotational motion is dominant. Section B is dominated by a phase gradient. This is the section in which the hot outgoing air mixes with the cool incoming air. We might call it the mixing length. This is the turbulent section of the pipe; the source of aberrations. At the same time, piston is much larger compared to other aberrations. Local piston has an average size of about 3.76 λ whereas the second most dominant one is defocus which has an average less than -0.005 λ, a factor of about 700. This implies that the density in each slice is almost uniform. From this, we can tell that the lens is very weak. Considering defocus, Fig. 10 (b), we notice immediately that focusing takes place in two parts of the pipe: the sections labelled B, with a large contribution from the region interfacing with section A, reaching a local maximum of -0.015 λ. Along its length, the SPGL has two centres of focus (sections B), thus making the SPGL very difficult to align. This confirms the 3D profile in Fig. 6 (a), that the lensing action of the SPGL comes from the mixing of hot and cold air.

The higher order aberrations increase dramatically in section B, suggesting that it is the mixing that gives rise to lensing also has a deleterious effect on the laser beam. This is because the mixing of hot and cold air creates local random varying density, which generates aberrations, the effect of which should increase with temperature and/or rotation speed. Spherical aberration reaches a local maximum of 0.0085 λ a magnitude half the size of defocus. Tilt also increases in the same region with a maximum of around 0.00002 λ. This behaviour is similar to the behaviour exhibited by coma (Fig. 10 (e)) though the values are about 10 times smaller.

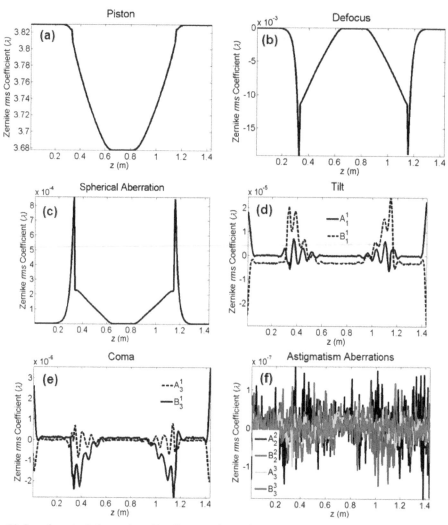

Fig. 10. Local optical aberration distribution along the SPGL calculated from the CFD density data

The beam quality factor distribution (Figure 11 (a)) in the SPGL confirms the aberration distribution. The beam quality factor is highest (suggesting a poor beam) at the same points where spherical aberration is highest. Of all the aberrations, spherical aberration has the largest coefficient. This confirms that spherical aberration is biggest contributor to beam quality deterioration in a gas lens. The wavefront error and Strehl ratio provide further proof of this. However, the important thing to note is that these parameters prove that the gas lens does not cause deterioration of the laser beam by that much. An unaberrated Gaussian beam has an M^2 of 1, whereas the model shows local values of M^2 of about 1.57. This results in a very low local wavefront error of about 0.0001 λ^2 on average. The Strehl ratio (Fig. 11b) is almost always 1 throughout the SPGL except in the mixing length where it drops by an infinitesimally small amount. The overall beam quality factor for the entire gas lens was found to be 2.5071 in both axes. The only disappointing aspect of the lens was its focal length which was found to be 5.03 m in both axes. This confirms that the gas lens is a very weak lens.

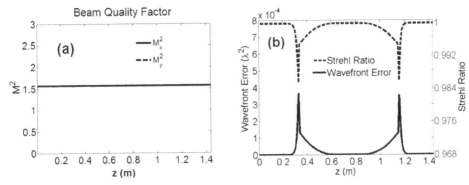

Fig. 11. The local beam quality factor (a), wavefront error and Strehl ratio (b) distributions along the pipe calculated from the SPGL CFD density data

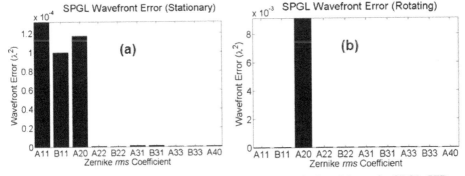

Fig. 12. Global primary optical aberrations except piston calculated from the SPGL CFD density data summarising the overall SPGL for a stationary (a) and rotating (b) heated pipe.

Now let us summarise the performance of the SPGL by looking at the CFD calculated global wavefront error for a heated stationary (Figure 12 (a)) and a rotating (Figure 12 (b)). The results for a stationary SPGL show the dominance of tilt in both axes, both with have values of at least 10^{-4} λ^2. Defocus has a value around the average of the two. This confirms that there is, indeed some focusing before rotation begins, though it is still very small, about 1.2×10^{-4} λ^2. However, as rotation commences, there is a significant increase in defocus, to 9.2×10^{-3} λ^2, which dominates other aberrations, including till, by a large amount. On the other hand, x- and y-tilt drop slightly from 1.3×10^{-4} λ^2 and 10^{-4} λ^2 to just below 10^{-5} λ^2. This confirms the fact that before rotation, the SPGL is dominated by tilt, due to gravitational distortion, but the effect is completely reduced under steady state rotation leaving defocus completely dominant. In other words, we have customised the density gradient to produce a lens.

3.5 Experimental verification of the SPGL model

Since density is directly proportional to refractive index, the phase of a laser beam propagating through a gas lens will be altered depending on the refractive index distribution. For completeness, we present a summary of the optical investigation on the aberrations generated by an actual SPGL that was characterised with a Shack-Hartmann wavefront sensor (Mafusire et al, 2008a). An expanded HeNe laser beam steered by flat mirrors is made to propagate through the lens. A Shack-Hartmann wavefront sensor was placed just behind the lens and used to measure the beam's quality and phase aberrations for rotation speeds up to about 17 Hz for wall temperatures 351, 373, 400 and 422 K.

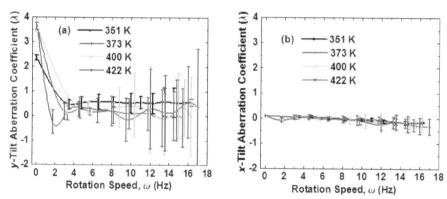

Fig. 13. y–Tilt (a) and x–tilt (b) generated by a spinning pipe gas lens at selected wall temperatures and rotation speeds.

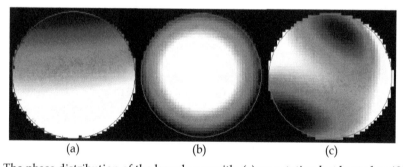

Fig. 14. The phase distribution of the laser beam with: (a) no rotation but heated to 422 K, showing tilt; (b) after rotating the SPGL at 17 Hz, showing significant curvature on the wavefront; and (c) same conditions as in (b) but with defocus and tilt removed, revealing the higher order aberrations.

The first result confirms the fact that rotation removes distortions which are caused by gravity: y-tilt, which is induced by gravity is reduced to a bare minimum as soon as rotation commences (Fig. 13 (a)) whereas x-tilt remains very small (Fig. 13 (b)) throughout. We can observe the same effect by looking the phase distribution before and during rotation. Figs. 14 (a) and (b) show the phase distribution before and during rotation, respectively. The phase maps are dominated by y-tilt (Fig. 14 (a)) before rotation, and defocus (Fig. 14 (b)) during rotation. However, digital removal of defocus and tilt reveals the presence of higher order aberrations, Fig. 14 (c). This phase map helps illustrate the other result observed during the experiment, the effect of the SPGL on the beam quality factor.

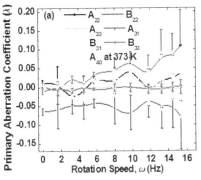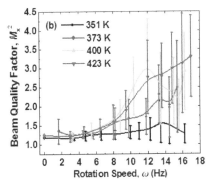

Fig. 15. (a) Higher order aberrations introduced by the SPGL; (b) increase in M_x^2 with rotation speed and temperature as a direct result of the aberrations in (a).

The other aberrations increase in magnitude as the rotation speed and/or wall temperature is increased (Fig. 15 (a)) thereby increasing the beam quality factor, M^2 (Fig. 15 (b)). This confirms the fact that the gas lens also generates aberrations which are increasing in power as the lens becomes stronger.

4. Optical turbulence

We now ask if the SPGL may be used as a controlled turbulence medium in the laboratory for the study of the propagation of optical fields through the atmosphere. The basis of this question is the fact that the SPGL introduces aberrations, and further that these aberrations may be controlled through the rotation speed and temperature of the pipe. Optical turbulence may also be described by aberrations, except that the weighting of the aberrations should take on a particular form. Characterising the aberrations in the SPGL from a turbulence perspective shows that the turbulence is uniform and isotropic near the pipe axis (about which it is spun), becoming non-uniform and anisotropic at the pipe boundary. A modified von Karman turbulence model (Andrews & Phillips, 2005) is used to analyse the turbulence strength along the pipe axis, and we find that the turbulence strength increases with rotation speed and pipe-wall temperature allowing for 'controlled' turbulence in the laboratory: our simple system allows for a controlled adjustment of the refractive index structure constant by more than two orders of magnitude.

The results of Fig. 16 illustrate this: in Fig. 16 (a) the log of the structure constant, which is a measure of the turbulence strength, is adjusted by two orders of magnitude as the pipe parameters of rotation speed and wall temperature are adjusted. Other supporting data (not shown here) confirms that the aberration weighting is correct for a particular atmospheric turbulence model – the modified von Karman turbulence model. This model has an inner and outer scale, the smallest and largest scales of the turbulence in the atmosphere respectively, relating directly to the smallest and largest fluid flow structures. These parameters can be measured in the pipe, and are shown in Fig. 16 (b). The pertinent point is that all the characteristics of the turbulence can be measured and therefore simulated in this simple aero-optic device, allowing for easy experiments of atmospheric turbulence in the laboratory.

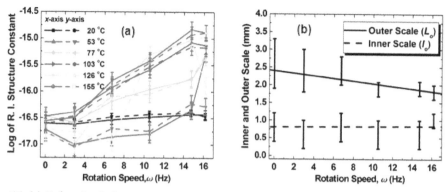

Fig. 16. (a) Refractive index structure constant at selected rotation speeds and wall temperature. (b) Inner and outer scales at the selected rotation speeds. These are standard parameters used to describe optical turbulence, and illustrate that our aero-optic may be used as a simulator of turbulence.

We also consider the optical aberrations imparted to the field when propagating near the boundary layer (Fig. 17), and find the phase distortions to the laser beam to be dominated by x-astigmatism with y-astigmatism and tilt, increasing dramatically in magnitude at the highest rotation speed and temperature. It is apparent that, in the spinning pipe gas lens, the parent flow is derived from the rotation of the pipe.

It is this rotation, together with the physical size of the lens, which limit the outer scale of the turbulence in the pipe. On the other hand, the outer scale is not much larger than the inner scale for a small inertial sub range, the range of scales between which the turbulence is isotropic, homogenous and independent of the parent flow.

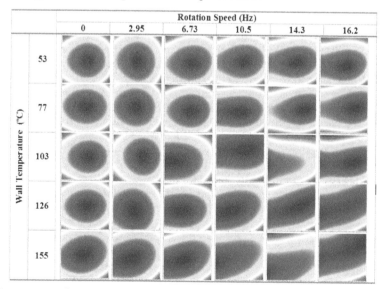

Fig. 17. Phase maps of the beam as it propagates near the boundary layer of the flow, for various wall temperatures and rotation speeds.

5. Summary

Aero-optics has found some novel applications of late, recently reviewed by Michaelis et al (Michaelis et al, 2006). These include long range telescoping elements, replacing high power laser windows to overcome damage threshold problems, adaptive lenses for delivery of high power laser beams in space propulsion experiments, and potential applications in laser fusion, control of peta-watt laser beams, photo-lithography with virtual capillaries and possibly novel guiding media for laser accelerators. Certainly if the quality of gas lenses could be improved, then the virtually limitless damage threshold of such lenses would make them ideal for most high power laser applications. The only drawback of most aero-optical devices is the distortions introduced to the laser beam due to imperfect control of the fluid, but as we have shown here, even this property may be exploited to simulate atmospheric turbulence in the laboratory.

In this chapter we have shown that it is possible to control the fluid flow inside a spinning heated pipe such that the density gradient of the air inside the pipe acts as a lens. As the focal length of this lens is a function of the rotation speed of the pipe and the temperature of the pipe wall, one has a variable focal length lens. We have shown focal lengths from infinity down to a couple of meters. We have shown that the lens is unfortunately aberrated, but highlighted that such aberrations un fact match atmospheric turbulence, so that the system may also be used as a simulator of atmospheric turbulence in the laboratory – again in a controlled and adjustable manner. It is this property – that such devices may be controlled – that makes aero-optics such an attractive possibility for future optical devices.

6. Acknowledgment

We would like to thank M.M. Michaelis for significant advice and for providing the original motivation for studying this field, and G. Snedden for his invaluable assistance in executing the CFD commercial code.

7. References

Andrews, L. C. & Phillips, R. L. (2005). *Laser Beam Propagation through Random Media*, SPIE Press.

Aoki, Y. & Suzuki, M. (1967). Imaging Property of a Gas Lens", *IEEE Trans. on Microwave Theory & Techniques*, Vol. 15, No. 1.

Blazek, J., (2001). *Computational Fluid Dynamics - Principles and Applications, Ch 7* Elvesier, Oxford.

Berreman, D. W., (1965). Convective Gas Light Guides or Lens Trains for Optical Beam Transmission, *J. Opt. Soc. Am*, Vol. 55, No. 3, pp. 239-247.

Born, M. & Wolf, E. (1998). *Principles of Optics: Electromagnetic theory of propagation, interference and diffraction of light 7th Ed* Cambridge University Press, Cambridge 517-553.

Dai, G-m, (2008). *Wavefront Optics for Vision Correction*, SPIE Press.

Davison, L., (2011). An Introduction to Turbulence Models. *Chalmers University of Technology.* Available from the University of Chalmers website: www.tfd.chalmers.se/~lada/postscript_files/kompendium_turb.pdf

Forbes, A. (1997). *Photothermal Refraction and Focusing*, PhD Thesis, University of Natal, Durban, South Africa.

Gloge, D. (1967). Deformation of Gas Lenses by Gravity, *Bell Sys. Tech. J.*, Vol. 46, No. 2, 357–365.

Kaiser, P. (1967). Measured Beam Deformations in a Guide Made of Tubular Gas Lenses, *Bell Sys. Tech. J.*, Vol. 47, 179–194.

Kaiser, P. (January, 1970). An improved thermal gas lens for optical beam waveguides, *Bell Sys, Tech. J.* Vol. 49, (January 1970) pp 137-153

Kellet, B.J.; Griffin, D.K.; Bingham R.; Campbell R.N.; Forbes, A. & Michaelis, M.M. (2008). Space polypropulsion, *Proc. SPIE* Vol. 7005, pp. 70052W-1.

Lisi, N.; Bucellato, R. & Michaelis, M. M. (1994). Optical quality and temperature profile of a spinning pipe gas lens, *Optics and Laser Technology*, Vol. 26, 25–27.

Mahajan, V. N., (1998). *Optical Imaging and Aberrations, Part I: Ray Geometrical Optics*, SPIE Press.

Mahajan, V. N. (2001). *Imaging and Aberrations, Part 2: Wave Diffraction Optics*, SPIE Press.

Mahajan, V. N., (2005). Strehl ratio of a Gaussian beam, *J. Opt. Soc. Am. A* Vol. 22, No. 9.

Martynenko, O. G., (1975). Aerothemooptics, *International J. of Heat and Mass Transfer*. 18, 793–796.

Michaelis, M. M.; Notcutt, M. & Cunningham, P. F. (1986). Drilling by a Gas Lens Focused Laser, *Opt. Comm.* Vol. 59, 369–374.

Michaelis, M. M.; Dempers, C. A.; Kosch, A. M.; Prause, A.; Notcutt M.; Cunningham, P. F. & Waltham, J., (1991). Gas lens telescopy, *Nature* Vol.353, 547-548.

Michaelis, M.M.; Forbes, A.; Conti, A.; Nativel, N.; Bencherif, H.; Bingham, R.; Kellett, B. & Govender, K. (2006). Non-solid, non-rigid optics for high power laser systems, *Proc. SPIE* Vol. 6261, pp. 15.1–15.13.

Mafusire, C. (2006). *Gas Lensing in a Heated Rotating Pipe*, MSc Thesis, University of Zimbabwe, (2006).

Mafusire, C.; Forbes, A.; Michaelis, M. M. & Snedden, G. (2007). Characterization of a spinning pipe gas lens using a Shack-Hartmann wavefront sensor", *Laser Beam Shaping VIII, Ed. F. Dickey, Proc. SPIE.*, Vol. 6663, 6663H.

Mafusire, C.; Forbes, A.; Michaelis, M. M. & Snedden, G. (2008a) Optical aberrations in a spinning pipe gas lens", *Opt. Exp.*, Vol. 16, No. 13, 9850-9856.

Mafusire, C.; Forbes, A.; Michaelis, M. M. & Snedden, G. (2008b). Spinning pipe gas lens revisited, *SA. J. Sci.*, Vol. 104.

Mafusire, C.; Forbes A.; Michaelis M.M. & Snedden, G. (2010a). Optical aberrations in gas lenses, *Proc. SPIE* 7789, pp. 778908-1.

Mafusire, C., Forbes, A. & Snedden, G. (2010b). A computational fluid dynamics model of a spinning pipe gas lens, *Proc. SPIE* 7789, pp. 77890Y-1.

Mafusire, C. & Forbes, A., (2011a). The Beam Quality Factor of Truncated Aberrated Gaussian Laser Beams, *J. Opt. Soc. Am. A*, Vol. 28, No. 7, 1372-1378.

Mafusire, C. & Forbes, A., (2011b). The Mean Focal Length of an Aberrated Lens, *J. Opt. Soc. Am. A* Vol. 28, No. 7, 1403-1409 (2011).

Mafusire, C. & Forbes, A., (2011c) Controlling optical turbulence in the laboratory, *Appl. Opt.* (submitted for publication).

Marcuse, D. (1965). Theory of a Thermal Gradient Gas Lens, *IEEE Trans. on Microwave Theory & Technology*, Vol. MMT-13, No. 6, 734–739 (1965).

Notcutt, M; Michaelis, M. M.; Cunningham, P. F. & Waltham, J. A. (1988). Spinning Pipe Gas Lens, *Optics and Laser Technology*, 20(5), 243–250.

Siegman, A. E. (1999). Laser Beams and Resonators: The 1960s, *IEEE Jour. of special topics in Quant. Elec.* Vol. 20, No. 5, 100–108.

Steier, W. H. (1965) Measurements on a Thermal Gradient Gas Lens, *IEEE Trans. on Microwave Theory & Technology*, Vol. MMT-13, No. 6, 740–748.

Zhao, Y. X., Yi, S. H., Tian, L. F., He, L & Cheng, Z. Y., (2010) An experimental study of aero-optical aberration and dithering of supersonic mixing layer via BOS, *Science China: Physics, Mechanics & Astronomy* Vol. 53, No. 1, 89-94.

11

Fluid-Dynamic Characterization and Efficiency Analysis in Plastic Separation of the Hydraulic Separator Multidune

Floriana La Marca, Monica Moroni and Antonio Cenedese
DICEA - Sapienza University of Rome, Rome
Italy

1. Introduction

Recovery of useable plastics from post-consumer and manufacturing waste remains a major recycling challenge. The global consumption of plastics was reported to be 230 million tonnes in 2005 (PlasticsEurope 2007a) of which 47.5 million tonnes were produced in Europe (25 European Union countries + Norway and Switzerland). Of the European production, only 22 million tonnes were reported as having been collected. Of this collected waste, 4 million tonnes were recycled as a manufacturing feedstock (18%) and 6.4 million tonnes went into energy recovery (29%), with the balance (11.6 million tonnes) probably being disposed in landfills (PlasticsEurope 2007b).

The recycling of plastics is a process essential to reduce the efflux of materials to landfills and to decrease the production of raw materials. In recent years awareness of the importance of environmental protection has led to the development of different techniques for plastic recycling. One issue related to the recycling of this material is the presence in the market of many types of plastics (polymers with additives), often with similar characteristics that make them difficult to differentiate in the recovery phase.

The separator "Multidune" is a hydraulic separator by density. Its name derives from the characteristic undulate profile of the channel where separation occurs. The channel is constructed from a sequence of closed parallel cylindrical tubes welded together in plane which are then sliced down the lateral mid-plane and the lower complex is laterally shifted relative to the upper complex. The Multidune allows solid particle separation according to their specific weight and the velocity field establishing within the apparatus.

Previous investigations (De Sena et al., 2008; Moroni et al. 2008) suggested the flow within the Multidune is organized into three main patterns. Principally, a longitudinal transport flow takes place, where the velocity is high. A particle belonging to this region can move from one camera to another. The second region is the lower recirculation zone with high values of the vorticity field. Particles belonging to this region undergo the vertical impulse of the fluid. The thrust is proportional to the vertical velocity component and, in conjunction with gravity and buoyancy, determines the destiny of a particle. If the thrust is larger than the net weight of the particle, an interaction with the principal transport flow occurs and, consequently, the particle will move to the following chamber. The third region is the upper recirculation zone whose dimensions are smaller than the other recirculation zone. If a particle moves from the principal flow to the secondary vorticity zone, it will have the chance to come back to the previous chamber, assuming the principal transport flow thrust does not prevent it from falling out.

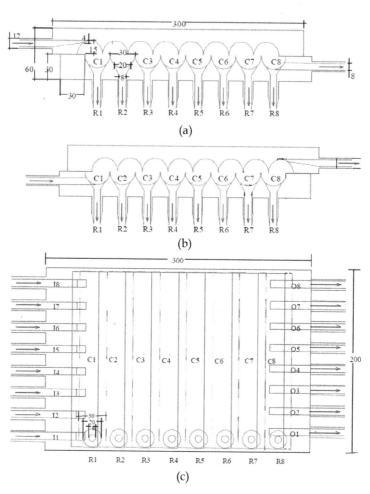

(a)

(b)

(c)

Fig. 1. Longitudinal section (dimensions in mm) for Arrangement (a) A and (b) B; (c) top view of the apparatus. C indicates the 8 chambers, R the 8 collection nozzles, I the 8 inlet nozzles and O the 8 outlet nozzles. Dimensions for Arrangement B are the same as for Arrangement A

Because of the role played by velocity, the fluid dynamic investigation of the Multidune apparatus is a preliminary step to carry out in order to investigate its capability in separating solid particles. For this reason a novel experimental campaign was set up and image analysis was employed to detect the velocity field within the apparatus. Two different arrangements of the Multidune were employed, hereinafter Arrangement A and

Arrangement B. Figure 1 presents a sketch of the longitudinal sections for the two arrangements, the top view and the apparatus dimensions in mm.

Two sets of experiments were run. The camera imaged all chambers of Arrangement A and B respectively and a passive tracer, i.e., a neutrally buoyant seeding particulate, was used to seed the fluid and reconstruct the fluid velocity field.

The fluid-dynamic investigation deals with the detection of several quantities of interest:

- mean velocity field and velocity profiles along the apparatus
- two-dimensional turbulent kinetic energy
- vorticity field

Generally, the detection of vortical structure (or coherent structure) immersed within a turbulent flow field is not straightforward. This book chapter will deal with this issue as well. Starting from the velocity gradient tensor, a topological analysis will be carried out with the aim of determining the critical points within the domain and to better represent areas with elevated values of the vorticity field.

The fluid-dynamic characterization will be correlated with the results of separation tests, carried out at lab-scale to separate particles of different plastic materials, characterized by low values in specific weight (about 1 g/cm^3, close to water specific weight).

2. Topological description of a fluid-dynamic field

It is usually difficult to individuate vortices, or coherent structures, within a statistical field which represents the background turbulent phenomenon. The detection of those structures is strictly related to their definition and it is influenced by the fact that the flow field can not be monitored at all scales. If an experimental monitoring technique is employed, a further limit is related to the availability of point, mono-dimensional, or two-dimensional measures.

Among the principal methods employed to detect vortices, visualization is the first and the currently most employed. The fluid motion can be visualized through the injection of dye or tracer particles. Remarkably, the surface where vorticity assumes a constant value does not necessarily match the surfaces where the dye concentration is constant. Even if the Schmidt number ($Sc=v/D$, where v is the kinematic viscosity and D the dye molecular diffusivity) is unitary, unlike the scalar behavior, vortices are subject to vortex stretching. Tracers would then present lower concentrations in areas of larger vorticity, and will not highlight fundamental behaviors of vortical dynamics. This inconvenient can be avoided seeding the fluid locally and detecting its behavior at distance and/or time intervals small respect to the characteristic scales (Joeng and Hussain, 1995).

Although the concept of vorticity is one of the most widely used in fluid mechanics, even today there is no generally accepted definition of vortex. A vortex is usually defined as the region of space enclosed by a surface formed by swirling lines (tangent at every point of the velocity curl). This definition is ambiguous as it would lead to identify a vortex within a laminar channel or in a Poiseille flow.

The conditions necessary for a proper definition of the vortex are (Hussain, 1986):

- inside the vortex, the vorticity ($\nabla \times u$) must assume non-zero values, it is a necessary but not sufficient condition;
- the existence of the vortex must be identified by a scalar quantity;
- identification criteria must be invariant under Galilean transformations.

Since the concept of vortex is associated with rotation of the fluid, some authors have proposed to identify a vortex where the pressure reaches the minimum required to balance the centripetal force. This definition falls in some situations, for example, in a convergent a minimum of pressure exists unless the fluid is characterized by a rotating motion.

Similarly, the use of trajectories and streamlines to identify a vortex does not appear appropriate since both are not invariant with respect to a Galilean transformation.

Some criteria for identifying vortical structures that meet the requirements above use scalar quantities derived from the velocity gradient tensor, ∇u or A_{ij} which is decomposed in its symmetric S_{ij} and antisymmetric W_{ij} parts. In particular, the eigenvalues of the velocity gradients are examined. In particular, the three criteria are based on:

- identification of regions where eigenvalues of ∇u are complex numbers
- identification of regions where the second eigenvalue of tensor $S_{ik}S+W_{ik}W_{ik}$ is negative (eigenvalues sorted in descending order)
- identification of the regions where the second invariant of the velocity gradient tensor is positive.

Those three criteria are equivalent in a two-dimensional case, as the case under investigation.

3. Experimental set-up

The Multidune apparatus, $0.30 \times 0.20 \times 0.06$ m^3 in size, is composed by a sequence of closed parallel cylindrical tubes welded together in plane. The device is sliced down its lateral mid-plane and the lower half is shifted laterally and then fixed relative to the upper half. Arrangements A and B (Figure 1) are shaped by two different positions of the lower half. In Arrangement A, the lower half is shifted 0.012 m right, while it is shifted 0.012 m left for Arrangement B. Each resulting chamber is labelled according to its position along the flow direction [C1 (first chamber), C2, ..., C8 (last chamber)]. The first chamber has eight round input nozzles (I1, I2, ..., I8) held at constant head ranging between 0.84 m and 2.69 m. The last chamber has 8 round output nozzles (O1, O2, ..., O8). Flow is induced in the lateral direction normal to the axis of the tubes by cutting inlet nozzles on one side of the device and outlet ones on the opposite side. Settled materials may be collected in the lower part of each half-cylinder.

The Multidune feeding occurs through a tank whose output pipe is split into 8 to distribute water within the apparatus through the 8 inlet nozzles. An overflow exit allows controlling the water level in the tank. The average flowrate within the Multidune apparatus will rely both on hydraulic head at the inlet nozzles and on the number of open outlet nozzles. With both arrangements, the experiments were run for three elevations of the tank (referred to the inlet nozzles height): 0.84 m, 1.84 m, 2.69 m. Three outlet nozzles (O2, O4 and O6) were open when Arrangement A was adopted. Table 1 reports the difference in fluid elevation between the tank and the middle of inlet nozzles and the corresponding flowrates. Preliminary tests conducted with Arrangement B suggested the flowrate determined by opening three outlet nozzles was too large to allow plastic particles separation. Then experiments have been run with only one opened outlet nozzle (O3). In this case, the flowrates establishing within the apparatus are consistently lower.

Experimental results of separation of both mono- and multi-material samples (Moroni et al. 2011; La Marca et al. 2011) suggest the behaviour of particles introduced within the Multidune apparatus strongly depends on the characteristic velocity within the apparatus, dimension and density of plastic particles. The geometry of the channel and the type of working fluid are considered to play a key role in the process of separation. We then introduce the non-dimensional parameters:

	Difference in fluid elevation between tank and Multidune	Average flowrate for Arrangement A	Re for Arrangement A	Average flowrate for Arrangement B	Re for Arrangement B
	(m)	(l/min)	-	(l/min)	-
Q1	0.84	9.35	584.38	4.57	380.58
Q3	1.84	11.82	738.75	5.00	416.83
Q5	2.69	13.53	845.63	6.55	545.58

Table 1. Difference in fluid elevation between the tank and the middle of the Multidune inlet nozzles and corresponding flowrates for both Arrangements.

- particle Froude number

$$\theta = \frac{V^2}{\frac{\rho_s - \rho}{\rho} g d} = \frac{V^2}{g' d} \tag{1}$$

where:
V: working fluid (water in the present case) characteristic velocity
ρ_s: particle density
ρ: water density
g: acceleration of gravity
d: particle diameter
g': buoyancy parameter
- Reynolds number

$$Re = \frac{V H}{\nu} \tag{2}$$

where:
H: apparatus characteristic dimension, set equal to the cylinder radius, i.e. 1.5 cm
ν: working fluid kinematic viscosity
Re for the three tank heights and both arrangements are reported in Table 1.
The following procedure was adopted to carry out experimental tests aimed at reconstructing the velocity field and separating the plastic materials:
1. set up of water tank height and hydraulic head;
2. water feeding in the apparatus through the 8 input nozzles;
3. saturation of the apparatus with water;
4. opening of the chosen output nozzles (O2, O4 and O6, in Arrangement A; O3 in Arrangement B);
5. passive tracer or sample feeding in the Multidune apparatus through the I3 e I4 input nozzles;
6. test execution, about three minutes;

7. only for separation tests, recovery of material expelled from the output nozzle(s);
8. closing of the output nozzle(s);
9. only for separation tests, recovery of material settled in each chamber;
10. weighting of recovered materials after 24-h drying.

4. Fluid-dynamic characterization by image analysis

The aim of the fluid-dynamic characterization of the Multidune apparatus was to reconstruct trajectories of tracer particles seeding the fluid under investigation and the velocity field evolution with time by means of image analysis techniques.

The need for measuring velocity fields has historically led researchers to develop experimental techniques and related instrumentation. An 'ideal' measurement system should be non intrusive to avoid flow field perturbations, should not require calibration, and should be suitable for obtaining the velocity field with a time and space resolution smaller than the characteristic time and length scales (i.e. Kolmogorov scale for turbulent flows). Velocity measurements based on optical methods capable of providing the velocity of tracer particles illuminated by a light source represent the best approximation of this 'ideal' system. The requirement is the working fluid to be seeded with neutrally buoyant particles which are assumed to follow the flow. The velocity vector is evaluated from the ratio of the tracer displacement, Δs, and the time interval Δt required for the displacement to take place. The time interval must be small enough for the approximation to be reasonable.

Digital images were acquired using a high-speed high-resolution (1280×1024 pixels) camera at a rate of 250 frames per second and stored for analysis. The camera axis was set perpendicular to the Multidune lateral face. A high powder lamp produces a light sheet for illuminating the interior of the channel. Green plastic power (mean diameter of about 200 μm) preconditioned with a solution of water and sodium hydroxide was used as the tracer for particle tracking. Preconditioning was used to neutralize the electrostatic charge on the particles. Tracer injection within the apparatus and data collection were started after the fluid reached steady state.

Feature Tracking (FT) was employed as the image analysis technique, i.e. a Particle Tracking algorithm which allows ignoring the constraint of low seeding density, being able to provide accurate displacement vectors even when the number of tracer particles within each image is very large (Moroni and Cenedese, 2005). FT reconstructs the displacement field by selecting image features (image portions suitable to be tracked because they remain almost unchanged for small time intervals) and tracking these from frame to frame. The matching measure used to follow a feature (and the L×H window around the feature, where L and H are the horizontal and vertical dimensions respectively) and its "most similar" region at the successive times is the "Sum of Squared Differences" (SSD) among intensity values: the displacement is defined as the one that minimizes the SSD. In Feature Tracking one applies the algorithm only to points where the solution for the displacement exists: those points are called "good features to track" (Shi and Tomasi, 1994). FT allows a Lagrangian description of the velocity field providing sparse velocity vectors with application points coincident with large luminosity intensity gradients (likely located along tracer particles boundaries). Lagrangian data are then used to reconstruct instantaneous and time-averaged Eulerian velocity fields through a resampling procedure.

Fig. 2a and Fig. 2b show the trajectories reconstructed by the FT algorithm within chambers C3 and C4 for experiments run with Arrangements A and B and difference in fluid elevation between the tank and the center of the Multidune inlet nozzles set to Q3.

Arrangement A	Arrangement B

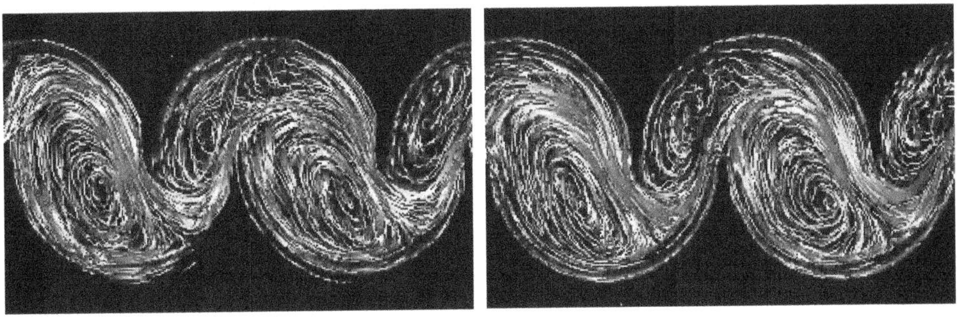

Fig. 2. Trajectories reconstructed by the Feature Tracking algorithm within chambers C3 and C4 for experiments run with Arrangement A (a) and Arrangement B (b) and difference in fluid elevation between the tank and the center of the Multidune inlet nozzles set to Q3

The trajectories are visualized overlapping 32 consecutive positions of the tracer particles. The colours range from blue (associated with the first time shown) to red (associated to the last one). The comparison between the two images evidences the increased length of the trajectories reconstructed when three nozzles are open instead of one. At the higher flowrate, the movement of tracer particles from one frame to the next is expected to be larger. The trajectories qualitatively describe the velocity field within the apparatus and confirm the presence of three sectors, i.e. the principal current which appears more consistent with Arrangement B, the lower recirculation area, larger with Arrangement A, and the upper recirculation area bigger with Arrangement B.

Figure 3 and Figure 4 present the velocity vectors overlapped to the colormap of, respectively, the horizontal and vertical velocity components within chambers C3 and C4, both arrangements and differences in fluid elevation between the tank and the center of the Multidune inlet nozzles set to Q1, Q3 and Q5.

The analysis of both velocity fields and streamlines (Figure 5) suggests the fluid-dynamic behaviour of the Multidune apparatus for both arrangements is characterized by three predominant areas.

The principal transport flow. It is characterized by a positive value of the velocity component along the x axis in the entire longitudinal section. The principal current is responsible for the transport of particles from one chamber to the next one and eventually it drives material to the outlet nozzles without separation. In Figure 5, it is indicated with red streamlines.

The lower recirculation zone. It is visible below the principal current, in each of the height chambers. The clock-wise rotating motion is suitable for subtracting particles from the principal current. It is expected that captured plastic particles will behave in one of the following ways:

- settle within the chamber if sufficiently heavy;
- follow the upward portion of the rotating motion without reaching again the principal current being too heavy to perform a complete rotation;
- execute a complete rotation and being captured again by the principal transport flow to settle in the next chamber or be expelled.

The particle behavior within the recirculation zone is influenced by its density and dimension as well as by the presence of a vortex. In Figure 5, it is indicated with green streamlines.

The upper recirculation zone. It develops behind the principal transport current. The purpose of the upper recirculation zone is to subtract particles from the principal current and to transfer them backward. To do so they have to pass across the principal current and to settle in the chamber. The particle physical attributes and the characteristic velocity of the principal current will influence the efficacy of this process. In Figure 5, it is indicated with blue streamlines.

The fluid vein characterizing the principal current is constituted by an inner part with velocity increasing with the hydraulic head at the apparatus inlet. Nevertheless, the characteristic dimension of the principal current remains practically constant even varying the flowrate. It appears more consistent with Arrangement B. At each flowrate, the upper recirculation zone developing in the Multidune- Arrangement A presents values of the both components of velocity field significantly lower than the characteristic velocity of the principal current. For this reason, this zone results ineffective in capturing particles from the principal current and then in the separation process. It was expected that reducing the flowrate and/or modifying the inner geometry of the apparatus, the upper recirculation zone would have played a role in the separation process. This is confirmed by the experiments with the Multidune- Arrangement B. Reducing the flowrate, the velocity of the principal current decreases whereas it remains practically constant within the recirculation areas becoming comparable. Furthermore, the modified geometry favours the formation of a larger upper recirculation area in each chamber incrementing the contact surface with the main current and the probability of material exchange between the two zones.

The velocity field in the lower recirculation area is in each case significantly lower than the characteristic velocity in the principal current. With the Multidune- Arrangement A, the velocity within the lower recirculation zone increases rising the flowrate, especially in the ascending portion of the vortex. This is actually counterproductive in terms of separation, because of the drag of settled particles toward the principal current. A similar behaviour with analogous consequences characterizes the descending portion of the upper recirculation area establishing in the Multidune- Arrangement B.

Noticeably, the analysis of the fluid-dynamic field developing within the Multidune apparatus with both arrangements suggests the increase of the hydraulic head augments the transport effectiveness of the main current without improving the capture feasibility of both the upper and lower recirculation zones. The apparatus will then lose its effectiveness in separating plastic particles increasing both the hydraulic head and the transiting flowrate. Furthermore, with both arrangements, C3 and C4 present analogous recirculation areas. The same circumstance occurs at each flowrate establishing in the apparatus. Furthermore, an analogous behavior characterizes C5 and C6. Then, the presence of eight chambers assures plastic particle separation even if a chamber should be filled with the settled material allowing the following chambers to become effective in the separation process.

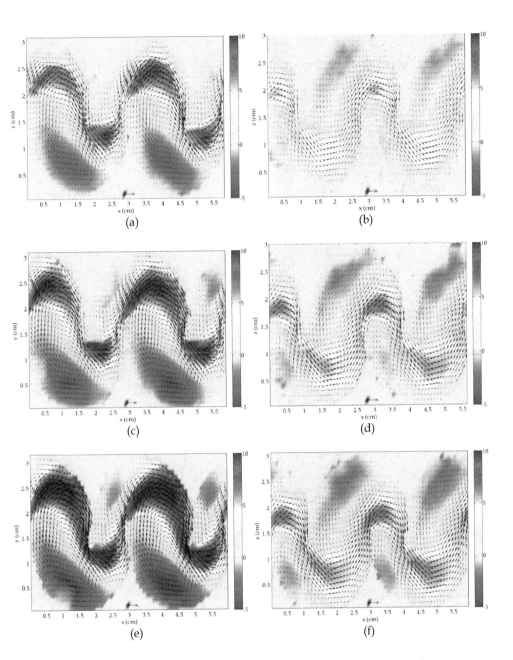

Fig. 3. Eulerian velocity field within C3 and C4 overlapped to the horizontal velocity component represented as colormap for Arrangement A at (a) Q1, (b) Q3, (c) Q5 and Arrangement B at (d) Q1, (e) Q3,(f) Q5. The reference velocity vector is equal to 10 cm/s

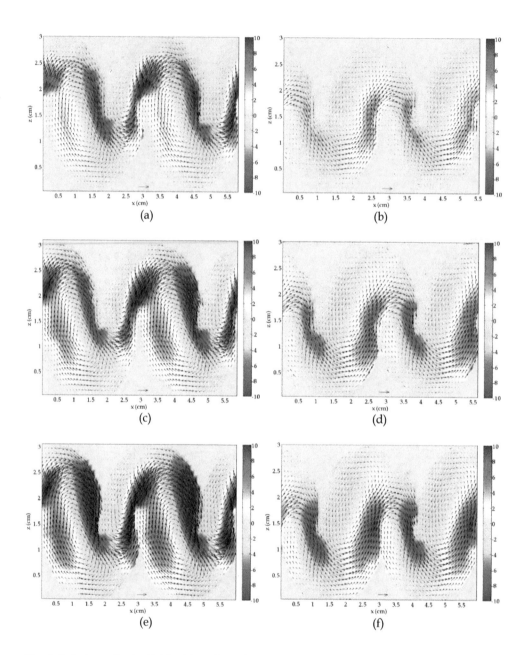

Fig. 4. Eulerian velocity field within C3 and C4 overlapped to the vertical velocity component represented as colormap for Arrangement A at (a) Q1, (b) Q3, (c) Q5 and Arrangement B at (d) Q1, (e) Q3,(f) Q5. The reference velocity vector is equal to 10 cm/s

Arrangement A   Arrangement B

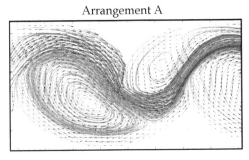 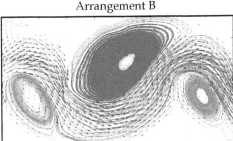

Fig. 5. Velocity field overlapped to the streamlines for Arrangement (a) A and (b) B. The difference in fluid elevation between the tank and the center of the Multidune inlet nozzles is set to Q3

Figure 6 presents the vorticity field computed for both Arrangements and for Q1, Q3 and Q5. The difficulty in identifying vortical structured by examining just $\nabla \times u$ should be clear from those figures. On the other hand, Figure 7 presents the complex part of the first eigenvalue of the velocity gradient tensor for both Arrangements and the difference in fluid elevation between the tank and the center of the Multidune inlet nozzles is set to Q5. It should be noted as the areas where the first (and the second as well, not shown) presents a complex value is within recirculation areas. It is then confirmed as the criteria previously mentioned for identifying vortical structures can be used for this experimental investigation.

5. Separation efficacy

Experimental tests were executed utilising four samples of plastics, in order to investigate efficiency and capability in differentiating trajectories according to plastic typologies and fluid-dynamic properties of the Multidune with both Arrangement A and B (Figure 1). The samples were composed with particles of different plastic material and size, as described in Table 2.
The tank heights and the corresponding flowrates utilized in the experimental tests were Q1, Q3 and Q5, as reported in the previous Table 1.

	Sample material	Mean density	Particle size distribution	Weight
Sample		(g/cm³)	(mm)	(g)
MONO.2	Brown plastic	1.416	0.85–1.00	1.5
MONO.3	Green plastic	1.353	0.85–1.00	1.5
MONO.4	Green plastic	1.353	1.70–2.00	2.5
MONO.5	Red plastic	1.143	1.70–2.00	2.5

Table 2. Composition of samples utilized in separation tests with Arrangements A and B.

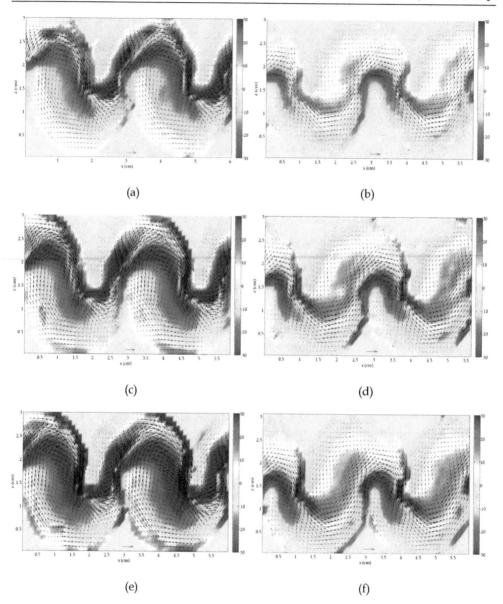

Fig. 6. Eulerian velocity field within C3 and C4 overlapped to the vorticity field represented as colormap for Arrangement A at (a) Q1, (b) Q3, (c) Q5 and Arrangement B at (d) Q1, (e) Q3, (f) Q5. The reference velocity vector is equal to 10 cm/s

The results of the tests on plastic samples suggested that the system may be able to separate different types of plastics in a mixture imposing an appropriate hydraulic head. In the following further remarks are given about the results referring to the adopted Arrangement.

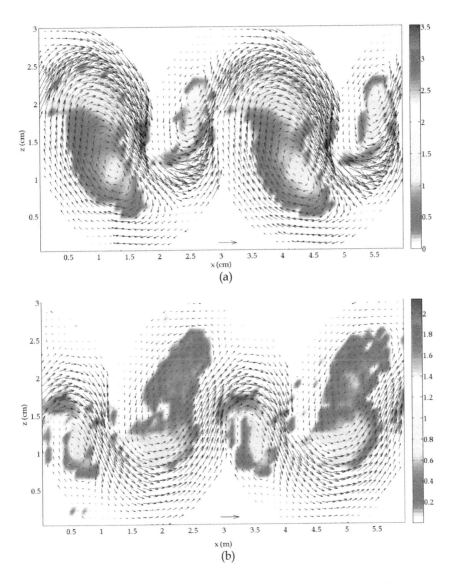

Fig. 7. Velocity field overlapped to the complex part of the first eigenvalue of the velocity gradient tensor for Arrangement (a) A and (b) B. The difference in fluid elevation between the tank and the center of the Multidune inlet nozzles is set to Q5

5.1 Results with arrangement A

Figure 8 shows the results of the experimental tests on plastic samples at each adopted flowrate, reporting the percentage in weight of the materials recovered from each chamber (C1-C8) and expelled through the output nozzles (Exp).

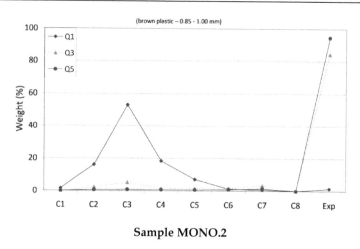

Sample MONO.2

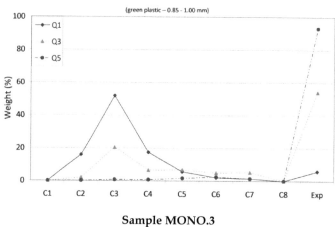

Sample MONO.3

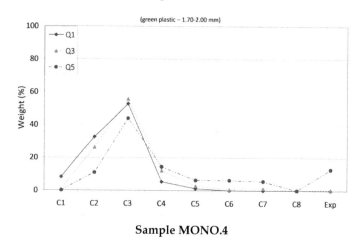

Sample MONO.4

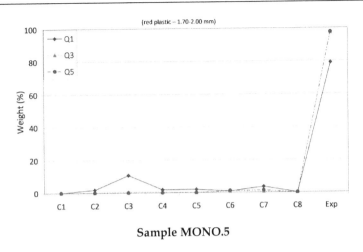

Sample MONO.5

Fig. 8. Results of the experimental tests on mono-material samples at each adopted hydraulic head with Arrangement A

It is evident as the highest hydraulic heads (Q3 and Q5) involve high flowrates, and, therefore, high velocities that sweep away all the small-size particles (0.85-1.00 mm, samples MONO.2 and MONO.3), without achieving satisfactory differences in respective trajectories.

On the other hand, with the lowest hydraulic head (Q1), the heaviest particles (samples MONO.2, MONO.3 and MONO.4) tend to settle into the apparatus, while the lightest ones (sample MONO.5) tend to be expelled through the output nozzle, regardless of particle size. In this case, a separation of the different types of plastics is possible, thanks to the differentiation of trajectories.

The more selective chamber is the third one, in which the greatest amount of settled material is accumulated, independently from applied flowrate. In the chambers C1 and C8, only few particles settled due to the presence of input and output nozzles, respectively.

Green and brown plastic particles sized 0.85-1.00 mm (samples MONO.2 and MONO.3) show the same behaviour, so a separation seems not possible for each tested operative condition.

Also in the case of coarser particles (samples MONO.4 and MONO.5), the particle settling in the chambers shows the same trend, being C3 the chamber in which mainly the plastic particles settle regardless of the applied hydraulic head, while C1 and C8 remain almost empty.

The green plastic particles sized 1.70-2.00 mm (sample MONO.4) tend to settle in the apparatus, with slight differences in the distribution in the chambers by increasing the hydraulic head; in particular, the particles shift from C2 to C3, and then from C3 to the following chambers.

The red plastic particles sized 1.70-2.00 mm (sample MONO.5) tend to be expelled from the Multidune. Only at the lowest hydraulic head, a significant amount of material settles in the C3.

Therefore, a different behaviour of the samples sized 1.70-2.00 mm for each tested hydraulic head was observed, so a separation between them seems possible to be achieved.

5.2 Results with arrangement B

Analogous experimental tests have been executed with the Arrangement B of the Multidune (Figure 1). As previously commented, the only difference in the operative procedure was the choice of the output nozzles to be opened. Firstly, the same three output nozzles as in Arrangement A were opened (O2, O4 and O6), but all the samples were expelled out because the high flowrate determined too high flow velocity, in each operative condition. Considering such results, the experiments were carried out by opening only one output nozzle (O3), to lower flowrate and velocity.

Figure 9 shows the results of the experimental tests at each adopted flowrate, in terms of percentage in weight of the materials recovered from each chamber (C1-C8) and expelled through the output nozzle (Exp).

In Arrangement B, it is confirmed as the small-size particles (0.85-1.00 mm, samples MONO.2 and MONO.3) are quite totally expelled from Multidune imposing the highest hydraulic heads (Q3 and Q5): so, in these operative conditions the trajectories of the two types of plastics are similar.

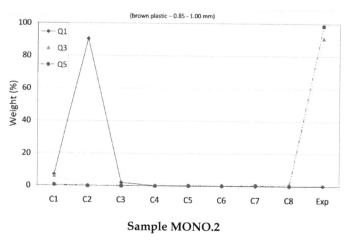

Sample MONO.2

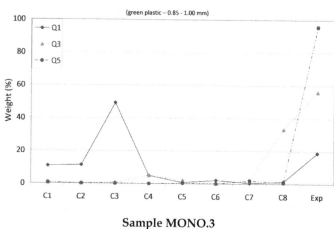

Sample MONO.3

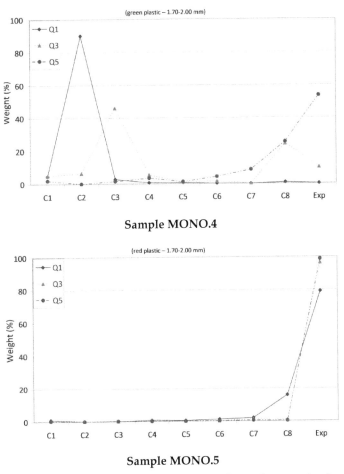

Sample MONO.4

Sample MONO.5

Fig. 9. Results of the experimental tests on mono-material samples at each adopted hydraulic head with Arrangement B

A separation seems to be possible at the lowest hydraulic head (Q1): actually, the heaviest particles (samples MONO.2, MONO.3 and MONO.4) tend to settle into the apparatus, while the lightest ones (sample MONO.5) tend to be expelled through the output nozzle, regardless of particle size.

The distribution of settled particles into the apparatus is more complex in Arrangement B than in Arrangement A. Particles path is in relation with the hydraulic head: as the hydraulic head increases, the particles tend to settled in the chambers closer to the output nozzles, consequently there is not a specific chamber (as C3 in Arrangement A) where particles are mainly accumulated.

The results of tests with Arrangement B confirm that it is possible to separate brown and green particles sized 0.85-1.00 mm (samples MONO.2 and MONO.3) imposing low flowrates, while green and red plastic particles of larger size (samples MONO.4 and MONO.5) can be separated at higher flowrates.

6. Conclusion

The high temporal and spatial resolution technologies employed for the reconstruction of the fluid-dynamic field inside the Multidune allows recognizing the flow field within the apparatus with both arrangements is characterized by three areas: the main transport current and, in each chamber, recirculation areas above and below. The fluid-dynamic behaviour is substantially similar in each chamber but in the first and the last one (C1 and C8), where the inlet and outlet nozzles prevent the formation of similar vortical structures. With Arrangement A, the characteristic velocity of the principal current appears significantly larger than the velocity within the upper and lower recirculation areas; this aspect is amplified with increasing hydraulic head at the apparatus inlet. With Arrangement B they appear more comparable. With both arrangements, the increase of the hydraulic head augments the transport effectiveness of the main current without improving the capture feasibility of both the upper and lower recirculation zones. The apparatus will then lose its effectiveness in separating plastic particles increasing both the hydraulic head and the transiting flowrate. Furthermore, with both arrangements, C3, C4, C5 and C6 present analogous recirculation areas. The same circumstance occurs at each flowrate establishing in the apparatus. Then, the presence of eight chambers assures plastic particle separation even if a chamber should be filled with the settled material allowing the following chambers to become effective in the separation process.

7. Acknowledgments

The authors would like to acknowledge Dr. Leonardo Cherubini and Dr. Emanuela Lupo for their contribution during the experiments.

8. References

De Sena G., Nardi C., Cenedese A., La Marca F., Massacci P., Moroni M. (2008). The Hydraulic Separator Multidune: Preliminary Tests on Fluid-Dynamic Features and Plastic Separation Feasibility. *Waste Management* 29(9), 1560-1571.

Hussain F. (1986). Coherent structures and turbulence, J. Fluid Mech. 173, 303.

Joeng G., Hussain F. (1995). On the identification of a vortex, J. Fluid Mech. 285, 69.

La Marca F., Moroni M., Cherubini L., Lupo E., Cenedese A. (2011). Recycling of plastic waste via the hydraulic separator Multidune. *Waste Management*, ISSN: 0956-053X, (submitted).

Moroni M., Cenedese A. (2005). Comparison among feature tracking and more consolidated velocimetry image analysis techniques in a fully developed turbulent channel flow. *Measurement Science and Technology* 16, 2307-2322.

Moroni M., Kleinfelter N., Cushman J.H. (2008). Alternative Measures of Dispersion Applied to Flow in a Convoluted Channel. *Advances in Water Resources* 32(5), 737-749.

Moroni M., La Marca F., Cherubini L., Cenedese A. (2011). Recovering plastics via the hydraulic separator Multidune: flow analysis and efficiency tests. *International Journal of Environmental Science and Technology*, ISSN: 1735-1472, (submitted).

PlasticsEurope (2007a). An analysis of plastics production, demand and recovery for 2005 in Europe. PlasticsEurope, 21 pp.

PlasticsEurope (2007b). Press Release 9 May 2007, 1 p. Association of Plastics Manufacturers in Europe (AISBL), Brussels, Belgium.

12

Optimization of Pouring Velocity for Aluminium Gravity Casting

Y. Kuriyama, K. Yano and S. Nishido
Gifu National College of Technology
Mie University
AISIN TAKAOKA CO., LTD
Japan

1. Introduction

In current casting factories, tilting type automatic pouring machines are often used to pour the molten metal into the mold, with the operator relying on experience, perception and repeated testing to manually determine the pouring velocity. However, seeking an optimum multistep pouring velocity through trial and error results in an enormous number of combinations and is very difficult. For this reason, it cannot be said that suitable casting that realizes a high-quality cast is being carried out, inviting a decline in the yield rate due to product defects.

Furthermore, the extension of the production preparatory phase and increase in costs due to this kind of trial operation also become a significant problem.

Until now, much research relating to product defects in aluminum gravity molding has been conducted (Yutaka et al., 2001)(Takuya, 2004). Meanwhile, research applying casting CAE for the purpose of improving the quality of castings and production efficiency is coming to attention (Itsuo, 2006). Furthermore, in recent years, optimization of the casting problem has begun to be carried out in accompaniment with developments in computers (Takuya et al., 2007)(Ken'ichi et al., 2008). However, these all target comparatively short calculation time die-casting and adoption of optimization technology in sand mold casting and gravity casting is lagging.

Accordingly, in this research, the objective is to stabilize the fluid speed in the mold gate and derive an optimum pouring velocity to realize a mitigation of defects such as pin holes and blow holes in aluminum gravity casting through invoking a fluid behavior simulator, swiftly filling the sprue cup and controlling the liquid level at a fixed high level of liquid. For the automatic pouring machine, the multistep velocity input is designed for actual product of an intake manifold. The effectiveness of this research is shown by a fluid analysis simulation and an actual test.

2. Experimental apparatus

An overview of the automatic pouring machine is shown in Fig. 1. It is a tilting type automatic pouring machine with one degree of freedom in the forward and back direction of the ladle.

In the pouring machine, the tilting angular velocity and velocity switching angle are configured by teaching pendant and the pouring velocity is determined. The setup enables four steps of tilting angle velocity and three of switching angle for a total of seven variables as shown in Fig.2. The tilting angular velocity command is assigned in the form of raised trapezoidal shapes as shown in Fig.3.

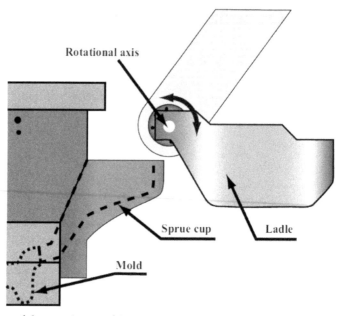

Fig. 1. Overview of the pouring machine

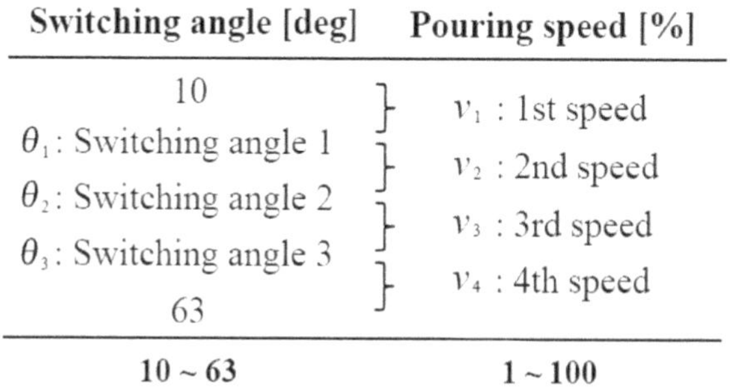

Switching angle [deg]		Pouring speed [%]
10	}	v_1 : 1st speed
θ_1 : Switching angle 1	}	v_2 : 2nd speed
θ_2 : Switching angle 2	}	v_3 : 3rd speed
θ_3 : Switching angle 3	}	v_4 : 4th speed
63		
10 ~ 63		1 ~ 100

Fig. 2. Input setting of the automatic pouring machine.

As the tilting angular velocity setting is displayed as a percentage, each step of tilting angular velocity is displayed by the following formula using the maximum angular velocity.

$$V_n = \frac{v_n}{100} V_{max} \tag{1}$$

Here, it is necessary to carry out parameter identification, since maximum angular velocity V_{max} and the angular acceleration a are unknown parameters.

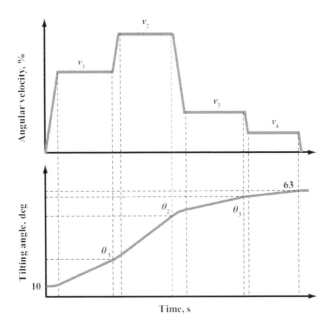

Fig. 3. Tilting velocity of pouring machine.

3. Identification the motion of the pouring machine

In this pouring machine, the tilting movement to input is unknown because the machine doesn't have the output device for the angular velocity or the angle. At the simulation of the fluid of molten metal, the identification of the tilting movement to input is needed. Thus, to get the unknown parameter of V_{max} and a, the movement to input of the actual poring machine is filmed with the motion capture system. Table 1 shows the input of the pouring machine using analysis the motion.

	Switching angle (deg)	Pouring velocity (%)
1	22	10
2	32	30
3	42	50
4	--	30

Table 1. Setting of the tilting input for identification the motion.

The unknown parameter of V_{max} and a, is identified by using the data of the angular velocity from the motion capture. The angular acceleration and maximum angular velocity are $a=200$ (deg/s²), $V_{max} =51.9$ (deg/s) respectively. Fig.4 shows the result of identification of actual pouring machine, where the solid line is the path of actual pouring machine and the broken line is the path of plant model.

From the Fig.4, it can be seen the motion is accorded very well. In this result, the plant model can be recreated the motion of the actual movement.

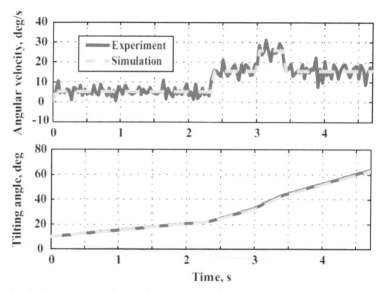

Fig. 4. Result of identification of actual pouring machine.

4. Construction of fluid behavior simulator and flow evaluation

The fluid analysis software FLOW-3D was used in this research. AC2B is taken as the subject fluid. The cast quantity is 1.863×10^{-3}(m³) and the product part volume is 1.429×10^{-3} (m³). The physical values of AC2B for the analysis were set as in Table 2. An outline map of the mesh in the simulation domain is shown in Fig. 5 and the mesh parameters are shown in Table 3.

Fluid parameters	AC2B aluminum alloy
Density	2550 kg/m²
Viscosity	0.00125 Pa·s
Temperature of fluid	993 K
Specific heat	1071 J/(kg·K)
Thermal conductivity	100 W/(m·K)

Table 2. Fluid parameters

Mesh block	Cell size (m)	Number of cell
X-direction	0.004~0.080	108
Y-direction	0.005	54
Z-direction	0.004~0.080	65
Total number of cell		379080

Table 3. Mesh parameters

In an actual plant, a molten metal filter (wire mesh) is installed in the sprue runner with the purpose of removing slag as shown in Fig. 5. The thickness of the wires in the mesh is 0.5×10^{-3}

(m) and the number of wires is 50 in the vertical and 55 in the horizontal. In this research a porous baffle (hereinafter, baffle) was used to reproduce this filter in the CFD simulator. The air porosity b_p, the linear velocity drop coefficient b_l and the two-dimensional velocity drop coefficient b_q are assigned as setting parameters for the baffle. b_l and b_q are defined by an equation for baffle flow loss shown by Eq. (2).

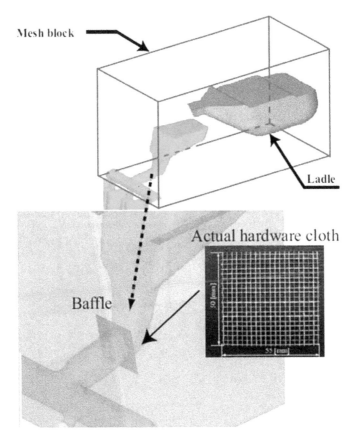

Fig. 5. Mesh setting of CFD

$$B = \frac{1}{L}\left(b_l u + 0.5 b_q u|u|\right) \tag{2}$$

Here, B denotes the baffle flow loss, u the flow speed within the baffle and L the length in which the flow loss is produced. The air porosity of the wire mesh can be calculated from the area ratio of the metal wires and opening between them. As the linear velocity drop is dominant for the baffle flow loss, the 2-D velocity drop coefficient is set at $b_q=0$ and the results of searching for the linear velocity drop coefficient b_l are shown in Table 4. Furthermore, the search range was carried out in 0.05 increments over $b_l=0.00$–1.50. Simulation results considering the baffle loss are shown in Fig. 6.

Void	0.655
Linear loss coefficient : b_l	0.90
Quadratic loss coefficient : b_q	0.00

Table 4. Parameters of porous baffle

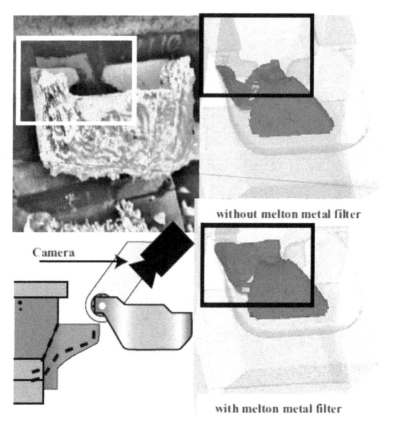

Fig. 6. Comparison of flow in sprue cup between simulation with molten metal filter and without molten metal filter.

Comparing these with the pouring test results, it can be seen that a satisfactory reproduction of the molten metal behavior inside the sprue is achieved.

5. Derivation of optimum pouring input using a genetic algorithm

Through swiftly filling the inside of the sprue cup with molten metal and controlling the liquid level to preserve a liquid level with high uniformity, the flow velocity in the mold gate is made constant.

This is regarded as making possible the production of high-quality casting that mitigates product defects. However, in the case the pouring velocity is determined by operator trial and error, problems occur such as the overflowing of molten metal from the melt due to

improper velocity setting and unstable liquid levels inside the sprue cup etc. Accordingly, in this research, liquid level control is realized through optimizing the pouring input using a genetic algorithm (GA) (Thomas et al., 1993).
GA is an algorithm that models natural selection and mutation in the processes of inheritance and evolution in biological groups in the processes of evolution and inheritance for populations in the natural world and is a probabilistic optimization method.
The three steps of switching angle and four steps of pouring velocity for a total of seven set parameters are taken as variables that are the settings input for the pouring machine, and an optimum tilting velocity pattern is calculated within the limitations of the real machine. In order to swiftly fill the sprue cup and stabilize the liquid surface at a uniform level, the tilting end time is taken as a performance function and the optimization problem is expressed by Eq. (3) with the liquid level inside the sprue as a limiting condition. Through taking the tilting end time as a performance function, the filling time is reduced and through already taking the liquid level as a limiting condition, liquid level control becomes possible.

$$\text{minimize} : J = t_p + J_p \tag{3}$$

$$h \geq 0.025$$

Here, t_p denotes the tilting end time, J_p a penalty function denoted by Eq. (4), and h the displacement from the sprue exterior to the molten metal.

$$J_p = w_1 + w_2 \tag{4}$$

In Eq. (4), the penalty clause $w_1,w_2=100$ is imposed in the case the displacement from the sprue exterior to the sprue interior liquid level drops to below 0.025(m) ($h < 0.025$).

6. Optimization with CFD simulator

6.1 Optimization result and melt flow analysis
Optimization by GA was carried out using the calculation parameters shown in Table 5 Forty-eight hours was required for optimization with 41st generation using an Intel Core2 Quad CPU equipped PC.

Number of variable	7
Number of population	10
Number of elite preservation	1
Mutation fraction	0.01
Crossover fraction	0.80

Table 5. Parameters of genetic algorithms

At that time the evaluation value was $J=4.668$. The tilting angles and velocities obtained from the optimum parameters are shown in Fig. 7. And the simulation results of the liquid level control is shown in Fig. 8.
As a result, it can be seen that the objective liquid level is not reached in the case of any control and a satisfactory liquid surface is not maintained. Conversely, it can be seen that a satisfactory liquid level control that swiftly fills the sprue cup is realized through the optimization of pouring control input.

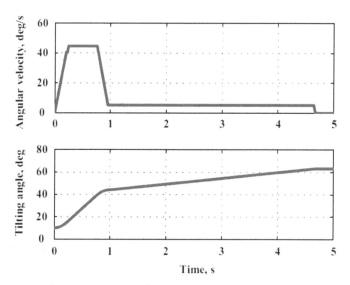

Fig. 7. Tilting input of optimization result.

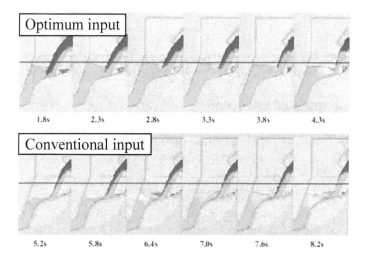

Fig. 8. Tilting input of optimization result.

6.2 Evaluation of the optimum input

Air entrainment is one of the defect origins of such as pin holes and blow holes as shown in Fig.9. Thus, using the evaluation function of air entrainment which is one of the functions of *Flow-3D*, the optimum input is evaluated.

The air entrainment at the liquid surface is based on the concept that turbulent eddies raise small liquid elements above the free surface that may trap air and carry it back into the body of the liquid. The extent to which liquid elements can be lifted above the free surface depends on whether or not the intensity of the turbulence is enough to overcome the surface-stabilizing forces of gravity and surface tension as shown in Fig. 10.

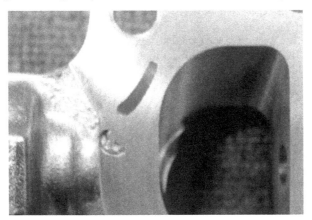

Fig. 9. Photograph of blow hole

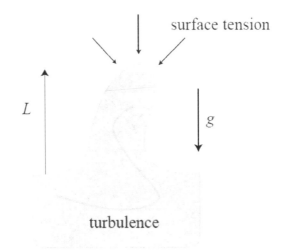

Fig. 10. Model of air entrainment

Turbulence transport models characterize turbulence by a specific turbulent kinetic energy Q and a dissipation function D. The characteristic size of turbulence eddies is then given by Eq. (5).

$$L = 0.1 \frac{\sqrt{Q^3}}{D} \tag{5}$$

This scale is used to characterize surface disturbances. The disturbance kinetic energy per unit volume (i.e., pressure) associated with a fluid element raised to a height L, and with surface tension energy based on a curvature of L is given by Eq. (6).

$$P_d = pgL + \frac{\sigma}{L} \tag{6}$$

where ρ is the liquid density, σ is the coefficient of surface tension, and g is the component of gravity normal to the free surface.

For air entrainment to occur, the turbulent kinetic energy per unit volume, $P_t=\rho Q$, must be larger than P_d; i.e., the turbulent disturbances must be large enough to overcome the surface-stabilizing forces. The volume of air entrained per unit time, V_a, is given as Eq.(7).

$$\frac{\partial V_a}{\partial_t} + \nabla V_a = Rt(1 - V_a) \tag{7}$$

where $R = C_{air}\sqrt{2(P_t - P_d)/\rho}$, u is fluid velocity, t is time, and C_{air} is a coefficient of proportionality : C_{air} =0.5; i.e., assume on average that air will be trapped over about half the surface area. If P_t is less than P_d then V_a is zero. Futhermore, Fig.11 is shown the meaasurement point of out flow rate and V_a.

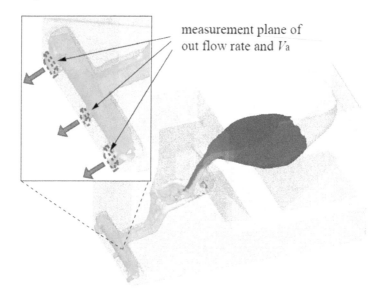

measurement plane of
out flow rate and Va

Fig. 11. Measurment plane of out flow rate and V_a

The air entrainment is expressed in Eq.(8).

$$A = \sum_{k=1}^{n} V_{ak}F_{fk}V_{fk}V_{ck} \tag{8}$$

where A is the quantity of air entrainment, V_a is the volume of air entrained per unit time, F_f is the fluid fraction, V_f is the volume fraction, V_c is the volume of the mesh cell, and n is the aggregate number of mesh cells.

Fig.12 shows the flow of molten metal with the mold. It can also be seen the satisfactory liquid level control that swiftly fills the sprue cup is realized. Fig.13 shows the air entrainment at the maching surface. Upper figure shows using the conventional input, lower figure shows the optimum input, where the color of this figure is indicated the value of the air entrapment, and red color is the most air entarinment point. From the figure, the optimum input can be decrease the air entrainment, and it can be expect to realize a mitigation of defects such as pin holes and blow holes in experiment.

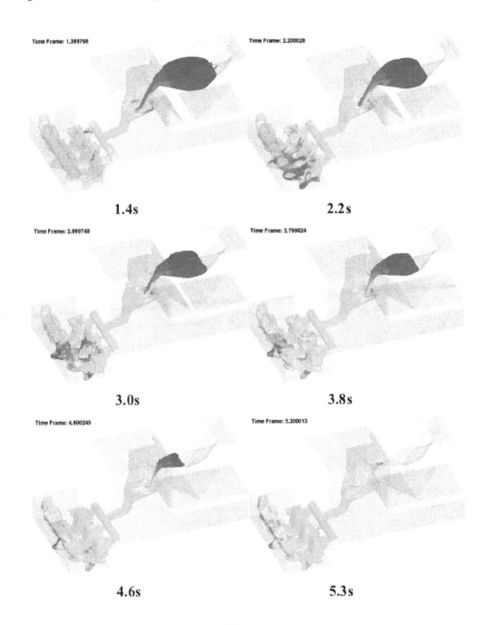

Fig. 12. Flow of molten metal with the mold

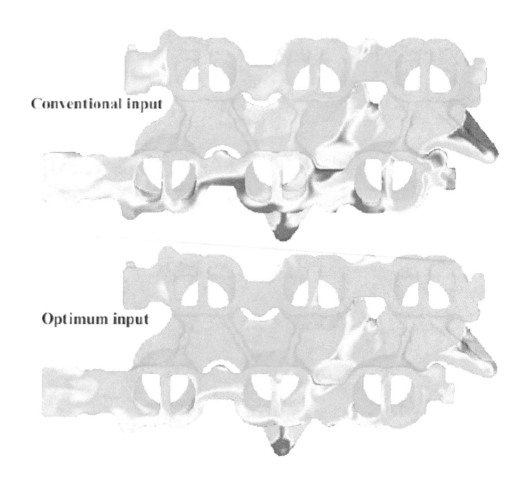

Conventional input

Optimum input

Fig. 13. Air entrainment at machining surface

7. Pouring control experiment using optimum velocity input

The desired optimum velocity input was applied to an actual machine and a pouring experiment was conducted. The conventional inputs were used to compare the optimal solution. Eleven trial runs were carried out and the frequency of defects appearing in the machined surface was evaluated after machining the product part from which the sprue, gating system and gate riser part had been removed. Fig. 14 shows the experiment results using each input and Fig. 15 shows the defect frequency for each.

As a result, it was possible to reduce the manufacturing defect to 1/6 rate by using an optimum pouring velocity as shown in Fig. 14 and Fig. 15. From this it was seen that the optimum input parameter derived using the proposed technique was valid for mitigating the occurrence of product defects.

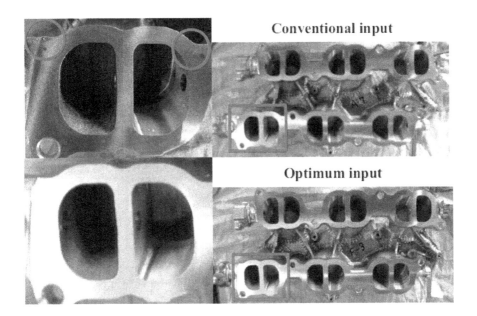

Fig. 14. Experimental results with conventional input and optimum input

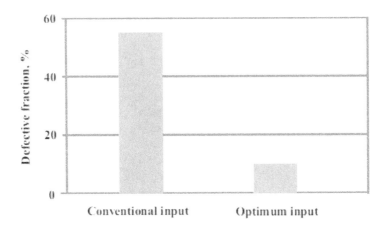

Fig. 15. Defect fraction of machining surface

8. Conclusion

In this research, an analysis technique that enables a reduction in calculation time in fluid analysis simulations was proposed and pouring control input was optimized with the purpose of reducing occurrence of defects such as blow holes and pin holes in aluminum gravity casting.

As a result of optimizing pouring control input using GA, optimum pouring control input realized liquid level control and its validity in mitigating product defects was seen through a real machine pouring experiment.

9. References

Yutaka K., Nobuaki N. & Hideaki O., (2001). Classification of Pin Hole Defect of Iron Casting. *Journal of Japan Foundry Engineering Society*, Vol.,73, No.,4, (2001 Apr.) pp. 258-263, 1342-0429

Takuya S.,(2004). Prediction of Gas Defects by Mold-Filling Simulation with Consideration of Surface Tension. *Journal of Japan Foundry Engineering Society*, Vol.,76, No.,7, (2004) pp. 562-569, 1342-0429

Itsuo O., (2006). State of the Art of Simulation of Casting. *Journal of Japan Foundry Engineering Society*, Vol.,78, No.,12, (2006 Dec.) pp.602-608, 1342-0429

Takuya S. & Ichiyo N., (2007). Optimization of Injection Speed for Reduction of Cold Shut in Die Casting, *Journal of Japan Foundry Engineering Society*, Vol.79, No.10,(2007 Oct.), pp.592-600, 1342-0429

Ken'ichi Y., Koutarou H., Yoshifumi K.& Seishi N., (2008). "Optimum Velocity Control of Die Casting Plunger Accounting for Air Entrapment and Shutting," *International Journal of Automation Technology*, Vol.2, No.4,(2008 Jun.) pp. 259-265, 1881-7629

Thomas B. & Hans- Paul S.,(1993). "An Overview of Evolutionary Algorithm for Parameter Optimization, " *Evolutionary Computation*, Vol.1, No.1,(1993 Apr.) pp. 1-23, 1063-6560

Permissions

The contributors of this book come from diverse backgrounds, making this book a truly international effort. This book will bring forth new frontiers with its revolutionizing research information and detailed analysis of the nascent developments around the world.

We would like to thank Dr. L. Hector Juarez, for lending his expertise to make the book truly unique. He has played a crucial role in the development of this book. Without his invaluable contribution this book wouldn't have been possible. He has made vital efforts to compile up to date information on the varied aspects of this subject to make this book a valuable addition to the collection of many professionals and students.

This book was conceptualized with the vision of imparting up-to-date information and advanced data in this field. To ensure the same, a matchless editorial board was set up. Every individual on the board went through rigorous rounds of assessment to prove their worth. After which they invested a large part of their time researching and compiling the most relevant data for our readers. Conferences and sessions were held from time to time between the editorial board and the contributing authors to present the data in the most comprehensible form. The editorial team has worked tirelessly to provide valuable and valid information to help people across the globe.

Every chapter published in this book has been scrutinized by our experts. Their significance has been extensively debated. The topics covered herein carry significant findings which will fuel the growth of the discipline. They may even be implemented as practical applications or may be referred to as a beginning point for another development. Chapters in this book were first published by InTech; hereby published with permission under the Creative Commons Attribution License or equivalent.

The editorial board has been involved in producing this book since its inception. They have spent rigorous hours researching and exploring the diverse topics which have resulted in the successful publishing of this book. They have passed on their knowledge of decades through this book. To expedite this challenging task, the publisher supported the team at every step. A small team of assistant editors was also appointed to further simplify the editing procedure and attain best results for the readers.

Our editorial team has been hand-picked from every corner of the world. Their multi-ethnicity adds dynamic inputs to the discussions which result in innovative outcomes. These outcomes are then further discussed with the researchers and contributors who give their valuable feedback and opinion regarding the same. The feedback is then collaborated with the researches and they are edited in a comprehensive manner to aid the understanding of the subject.

Apart from the editorial board, the designing team has also invested a significant amount of their time in understanding the subject and creating the most relevant covers. They scrutinized every image to scout for the most suitable representation of the subject and create an appropriate cover for the book.

The publishing team has been involved in this book since its early stages. They were actively engaged in every process, be it collecting the data, connecting with the contributors or procuring relevant information. The team has been an ardent support to the editorial, designing and production team. Their endless efforts to recruit the best for this project, has resulted in the accomplishment of this book. They are a veteran in the field of academics and their pool of knowledge is as vast as their experience in printing. Their expertise and guidance has proved useful at every step. Their uncompromising quality standards have made this book an exceptional effort. Their encouragement from time to time has been an inspiration for everyone.

The publisher and the editorial board hope that this book will prove to be a valuable piece of knowledge for researchers, students, practitioners and scholars across the globe.

List of Contributors

Sujudran Balachandran
Bumi Armada Berhad, Malaysia/Singapore

Katsuya Nagayama and Keisuke Honda
Kyushu Institute of Technology, Hitachi Cooperation, Japan

Daniel A. Marinho
University of Beira Interior, Department of Sport Sciences, Covilhã, Portugal
Research Centre in Sports, Health and Human Development, Vila Real, Portugal

Tiago M. Barbosa
Research Centre in Sports, Health and Human Development, Vila Real, Portugal
Polytechnic Institute of Bragança, Department of Sport Sciences, Bragança, Portugal

Vishveshwar R. Mantha and António J. Silva
Research Centre in Sports, Health and Human Development, Vila Real, Portugal
University of Trás-os-Montes and Alto Douro, Department of Sport Sciences, Exercise and Health, Vila Real, Portugal

Abel I. Rouboa
Research Centre in Sports, Health and Human Development, Vila Real, Portugal
University of Trás-os-Montes and Alto Douro, Department of Engineering, Vila Real, Portugal

Renat A. Sultanov and Dennis Guster
Department of Information Systems and BCRL, St. Cloud State University, St. Cloud, MN, USA

Vinicius C. Rispoli and Joao L. A. Carvalho
Universidade de Brasília, Brazil

Jon F. Nielsen
University of Michigan, USA

Krishna S. Nayak
University of Southern California, USA

Kleiber Bessa
Department of Environmental Sciences and Technological, Rural Federal University of Semi-Arid, Brazil

Daniel Legendre and Akash Prakasan
Institute Dante Pazzanese of Cardiology, Brazil

Takahisa Yamamoto
Gifu National College of Technology, Japan

Seiichi Nakata
Fujita Health University, Japan

Tsutomu Nakashima
Nagoya University, Japan

Tsuyoshi Yamamoto
Kyushu University, Japan

Marco Marcon
Politecnico di Milano, Dipartimento di Elettronica e Informazione, via Ponzio 34/5, 20133Milano, Italy

H. Pérez-de-Tejada
Institute of Geophysics, UNAM, Mexico

Cosmas Mafusire and Andrew Forbes
Council for Industrial Research National Laser Centre, South Africa
School of Physics, University of KwaZulu-Natal, South Africa

Floriana La Marca, Monica Moroni and Antonio Cenedese
DICEA - Sapienza University of Rome, Rome, Italy

Y. Kuriyama, K. Yano and S. Nishido
Gifu National College of Technology, Mie University, AISIN TAKAOKA CO., LTD, Japan

Printed in the USA
CPSIA information can be obtained
at www.ICGtesting.com
JSHW011414221024
72173JS00004B/543